LONDON BOROUGH OF ENFIELD
LIBRARY SERVICES

This book to be RETURNED on or before the latest date stamped
unless a renewal has been obtained by personal call or post,
quoting the above number and the date due for return.

WARGAMING
Ancient and Medieval

WARGAMING
Ancient and Medieval

Donald Featherstone
Maps by Catherine Stuart

DAVID & CHARLES

NEWTON ABBOT   LONDON   VANCOUVER

HIPPOCRENE BOOKS INC   NEW YORK

HIPPOCRENE
BOOKS, INC.

This edition first published in 1975 in Great Britain by
David & Charles (Holdings) Limited, Newton Abbot, Devon, and in
the United States of America in 1975 by Hippocrene Books Inc.,
New York

Published in Canada by Douglas, David & Charles Limited,
132 Philip Avenue, North Vancouver BC

ISBN 0 7153 6939 3 Great Britain
ISBN 0-88254-353-9 United States

Library of Congress Catalog Card Number: 75-7009

Set in Times New Roman
and printed in Great Britain
by Latimer Trend & Company Ltd Plymouth

Contents

Contents

List of Maps

Introduction

A WARGAMER specialising in the ancient and medieval periods might be attracted by the prospect of reconstructing a battle such as Cannae (fought on the equivalent of 3 August 216 BC), which was Hannibal's greatest victory over the Romans and a model of tactical perfection. Unfortunately, because of the large numbers involved, it is almost impossible to put this into practice, for the Roman force consisted of 8 Roman and 8 allied legions, totalling 80,000 infantry with 8,000 cavalry, while Hannibal's Carthaginian army had 40,000 infantry and 10,000 cavalry. Even scaling down so that 100 men on the battlefield equal only 1 man on the wargames table still calls for 1,200 infantry and 180 cavalry, and a very large wargames table will be required if these numbers are not to be so packed together as to make manoeuvre impossible. Much the same problems exist with such other battles as Zama, 202 BC; Trebbia, 218 BC; Hydaspes, 326 BC; Chalons, AD 451; and Hastings in 1066. However, numerous smaller battles and skirmishes are suitable for reproduction on the wargames table, the forces taking part and the topographical features being easy to simulate.

All the battles in this book possess tactical and human aspects worthy of simulation, and a third of them include significant and outstanding examples of the art of warfare and the development of tactics. Fought in 1288 BC, the Battle of Kadesh was not only one of the earliest recorded battles but is almost certainly the first-known instance of a tactical outflanking movement; the Battle of Leuctra, 371 BC, demonstrates the earliest example of attack in the 'oblique order' later used by eminent commanders such as Frederick the Great in the mid-eighteenth century; Alexander's use of 'artillery' at the River Jaxartes in 330 BC was not paralleled for centuries; Cynoscephalae in 197 BC was the start of the triumphal march of the Roman legions, heralding the decline of the historic phalanx, although this formation was successfully resuscitated by the Swiss mercenaries some 1600 years later; and the defeat of Harald Hardrada and his Norsemen at Stamford Bridge in 1066 put an end to centuries of Scandinavian invasions of the British coast.

Some of the best-known battles of the medieval period do not lend themselves to reproduction because of their lack of outstanding tactical features. Most engagements of this period were nothing more than scuffles of scrambling men and horses over a patch of bare land or a hillside, as armies formed themselves into great masses (battles) to be independently launched at the nearest enemy; lack of discipline and control ruled out any combined tactical movements.

Numbers, terrain and dearth of information eliminate many interesting battles. Further battles from the Wars of the Roses, and battles involving Swiss mercenaries or Ottoman Turks, have been omitted because artillery powered by gunpowder or handguns were used in them, and we are not including such engagements. This decision has led to the omission of Formigny (1450), the penultimate battle of the Hundred Years War and a small but fascinating engagement that decided the fate of all Normandy, where a pair of culverins were charged and captured by English archers. Not included because of lack of detailed battle reports and maps, or forces too large for realistic scaling down, are wars involving Assyrians or Persians, and other ancient empires, the Romans in Britain, the Crusades, the Moors in Europe and the Mongol or Hun cavalry armies. Many small battles, otherwise ideal for our purpose, lack maps, among them a very stimulating Roman/Carthaginian engagement at Tunis in 255 BC, and another near Carthage in AD 18, which is described by Robert Graves in his novel *I Claudius*.

The wargamer successfully refighting the battles herein needs to utilise the Bibliography and read all the available accounts of the action and its background. In so doing he will become aware of a commander's intentions and the tactical factors that brought victory to one side and defeat to the other. Our brief battle accounts, however, consider each tactical phase and, in the form of Military Possibilities, suggest possible alternatives, to provide a guide to accurate and realistic reconstructions.

In the battles described in this book the numbers involved are never too great for table-top scaling down to a ratio of 1:20. Thus at Leuctra the Thebans' 6,000 men will be represented by 300, and the Spartans' 10,000 by 500 figures. In those battles where only part of an army was engaged, or its separate parts came into action at varying times, the same group of figures can be used more than once, drastically reducing the number of model soldiers to be bought and painted.

Reconstructing Battles
Except when being purposefully staged as a demonstration, wargaming is a competitive affair, and the battles described in this book present the challenge of reversing their result. The table-top terrain over which they are fought must closely resemble, both in scale and appearance, the original battlefield, and the troops, though scaled down in numbers, must accurately represent the original forces. The wargamers, while following the original course of events, are allowed some leeway in the form of Military Possibilities, but these must be reasonable alternatives, or the reconstruction will lose its authenticity and become an ordinary wargame.

If the battles are so reconstructed that both armies are of historically accurate scaled-down strength, pursue the same tactical plans, use the same weapons and fight in a manner of their day, it is highly likely that the table-top encounter will follow its historical course. If not, the rules controlling the wargame may lack balance, or one wargamer, perhaps better versed in military tactics than his opponent, is manoeuvring his armies with a military hindsight denied to the real-life commander of long ago. Wargames armies must not be tactically manoeuvred in a manner far beyond the knowledge and comprehension of their time. Such restraint is difficult for the twentieth-century wargamer, with his knowledge of later military actions, and his awareness of the historical course of the battle.

The wargamer handling the losers may feel disinclined to follow a course of action so obviously doomed to failure, although he must realise that the stereotyped course taken by the losing commander was probably that prescribed by the military thought of the day. The military mind does not learn quickly – it took the French most of the Hundred Years War to revise the tactics that led to their repeated defeats by the English archer and his longbow.

A number of factors, frequently considered in the various battle reconstructions, now require more detailed explanation.

Commanders' Classification

Two factors common to the majority of these battles are that the victorious commander, usually with a numerically weaker army, possessed outstanding tactical ability and was capable of inspiring his men to fight 'above themselves'. Notable instances are Epaminondas at Leuctra, Alexander at the River Jaxartes, de Montfort at Lewes, Bruce at Bannockburn, Harold at Stamford Bridge, Northampton at Morlaix and Derby at Auberoche. In some of the other battles the victorious commander just happened to be better than his opposite number, as at Shrewsbury and Neville's Cross. Then there are the 'soldiers' battles', where victory depends not on the commander but purely on the prowess of his men, as at Verneuil. Finally, subordinate commanders sometimes merit higher ratings than their leaders, as at Cynoscephalae, where the course of the battle was dramatically changed by the inspired move of an anonymous Roman Tribune.

The difference in commanders' ability has to be simulated on the wargames table so that those troops led by an Above Average commander display higher morale and better fighting qualities than the unfortunates led by a Below Average commander. There are numerous ways of reflecting this, but the simplest way is to add or subtract pre-decided numbers to all dice scores reflecting morale and fighting ability. The scores of troops led by an 'average' commander remain unaltered.

Commanders of varying classifications are granted or denied

certain tactical qualities. An Above Average commander is allowed flexibility of movement in that he is permitted to handle his force in a tactical manner outside his period, the Below Average commander is compelled to manoeuvre and control his army in the strict manner of their 'Style of Fighting', and the Average commander is given some freedom to alter characteristic tactical and combat factors but must control and manoeuvre his force in accordance with their 'Style of Fighting'.

Their awareness of the tactical course of events in the battles under review allows wargames commanders to issue pre-battle orders, in the manner decreed by the rules of the Wargames Research Group. Under these circumstances the Below Average commander has no alternative but to stick rigidly to his initial intentions until they are rendered inoperable by the ebb and flow of the battle. A force he has ordered to move up a hill and occupy its crest will do so even if it becomes obvious that this can only end in disaster. An Average commander is permitted to change his orders at the conclusion of a prearranged number of moves, say 3, because of changed circumstances. The Above Average commander is permitted to write fresh orders at the beginning of each move of the game. The grading of commanders can also be simulated by selecting novice wargamers as Below Average commanders and veteran wargamers as Above Average.

Military Possibilities

Each battle account includes a number of credible contingencies, known as Military Possibilities, which can produce uncertainty and even reverse the eventual result. Military Possibilities can sometimes produce a more credible outcome than that revealed by history! For example, it is not unreasonable to assert that if the Norsemen of Harald Hardrada, being matched only by Harold's Housecarls, had had time to don their mail and unite, they would probably have won at Stamford Bridge. The course of history might then have been dramatically altered, for they and not Harold would later have confronted William of Normandy and his invading army. This is a Military Possibility

based on fact, but possibilities can be devised to represent the inevitable fluctuations of fortune, and invoked through the throw of a dice or the turn of a Chance Card.

Military Possibilities bring to the wargames table a certain spice and colour, and, in proportion to the wargamer's ingenuity, they can radically alter the historical course of a battle or be so tempered as to affect some relatively minor aspect only. They are used in this book primarily as a means of encouraging the wargamer representing the defeated commander to accept his role by giving him an outside chance of reversing the result.

Military Possibilities pose some fascinating questions. If Philip had refused battle at Cynoscephalae on terrain unsuitable to the Macedonian phalanx, would the brilliance of the Roman legion have been stillborn? If Prince Edward's cavalry at Lewes in 1264 had not prematurely emulated Rupert's horsemen at Edge Hill some 400 years later by pursuing their beaten enemy instead of playing a decisive role in the battle, how would the course of English history have been changed? Would Henry V have been there to triumph at Agincourt in 1415 if Owen Glendower had arrived to strengthen Hotspur's army at Shrewsbury in 1403? If the archer whose arrow scarred Prince Hal's face had aimed a little better, he would certainly not have been.

Chance Cards

Representing the 'human' influence on the tactical aspects of a Military Possibility, Chance Cards force a commander to take practical steps to implement their instructions, and his actions in so doing may materially affect the course of a battle. Among the battles under review, Chance Cards could determine whether or not the Hittites plundered the Egyptian camp at Kadesh by detailing (a) whether plundering took place or not; and (b) if it did, for how many game moves. The speed at which the Norsemen donned their mail or the speed at which their reinforcements arrived at Stamford Bridge could both be detailed in game moves representing time. Chance Cards could postulate a shortage of arrows for the English longbows at Morlaix, by setting out the number of arrows held by each archer and stating whether or not further supplies are to hand. They could allow Hotspur not to be killed at Shrewsbury, or give him, like the King, 'doubles' wearing his heraldic devices. They could order the Lombard mercenary mounted archers to ignore the wagons at Verneuil. Many more chance actions may be devised by the wargamer, for every battle presents opportunities to the imagination.

Time Charts

Kadesh, the very first battle in the book, has its 'surprise' factor, concerning the time of arrival of reinforcements; and they must arrive on the wargames table at the appropriate period in the battle. Such vital aspects must be programmed on a Time Chart to ensure their occurring in the correct sequence. 'On the table' moves, such as the 20 maniples attacking the Macedonian right at Cynoscephalae, Derby's synchronised attack at Auberoche, and Warwick's advance through the houses at St Albans, also require recording on a Time Chart. In the same manner a check must be kept on 'off-table' moves, such as the Norse reinforcements coming from Riccall during the Battle of Stamford Bridge, and Prince Edward and his cavalry pursuing the enemy from the field at Lewes. Any situation where forces are attempting a manoeuvre that will bring them on to the table-top battlefield at some intermediate stage in the conflict requires recording on a Time Chart. If, as at Stamford Bridge, a commander sends messengers imploring aid, then the progress of these messengers, their exact time of arrival, and the subsequent time reaction of the unit receiving the message, must be recorded on a Time Chart, as must the progress of a relieving force.

As a practical example of the use of a Time Chart, consider Rameses leading his 4 divisions of the Egyptian army at Kadesh. After being shown a ford over the river near Shabtuna, the Egyptian vanguard pushed ahead of the army as each division halted to water and feed at the ford. After crossing a ford the head of a column should halt on the far bank to allow the remainder to close up, if the army is not to be dangerously strung out along the road. Such halts by each division will cause a gap between them, and their progress must be clearly shown on a Time Chart. The movements of the Vizier sent by Rameses to hasten the march of the 2 rear divisions must be marked on the Time Chart, as must the delay caused to the Hittites by plundering the Egyptian camp. The time when the Vizier came up with troops emerging from the Forest of Baui must be recorded, together with the time of his meeting the Division of Ptah. Finally, the time of arrival of the Egyptian reinforcements has to be noted. All these timings are indicated by numbered game moves.

Surprise
Because of the omniscience of the wargamer, the factor of 'surprise' is extremely difficult to simulate on the wargames table. When reconstructing a historical battle in which there was a surprise factor, this element must be represented if the conflict is to be accurately reproduced. A certain degree of tactical surprise and concealment is possible when movements are charted on a map, forces being manoeuvred without the enemy being aware of their intention, or even their existence, until contact is made.

Another method of achieving surprise, even with forces in full view on the table top, is for each commander initially to draw up a plan of the tactics (down to unit level) he intends to use. Then 8 playing cards, 2 of them aces, are dealt out to each commander. Before each game move, he draws a card and, if it is an ace, he is permitted to alter the allotted role of a unit. In circumstances where history indicates one force (or its leader) to be markedly superior, that force can be awarded a higher proportion of aces than its opponent. Using this method, the complete surprise achieved by Derby at Auberoche can be simulated by his being given 4 aces in his 8 cards, whereas the French are allowed 1 at the most.

Of the 15 battles described in this book, the element of surprise was present at Kadesh, Leuctra (the Theban formation), Stamford Bridge, Auberoche and St Albans (Warwick's centre attack). But all these battles may produce surprises for the wargamer by their unfamiliarity. For example, if he is commanding the force of Scythian horse archers who are attempting to prevent Alexander's Macedonians crossing the River Jaxartes, he may well be as astonished as was the original Scythian commander by Alexander's brilliant employment of 'artillery', which laid down covering fire while the Macedonian infantry crossed the river. In addition, lulled into a false sense of security by his knowledge of the great success of horse archers at Carrhae and on other fields, the 'Scythian' wargamer may be also quite unprepared for Alexander's unconventional cavalry tactics that rendered them impotent at Jaxartes.

Another way of manufacturing surprise is for the host wargamer not to give the visiting player the name of the battle he is fighting. With Auberoche, for instance, the information that a divided French force is besieging a small English force in a castle can be all that is provided. The besiegers are arranged in their two camps, and as the battle proceeds, the visiting 'French' commander will no doubt be just as surprised as was the original French leader when he is outwitted by a much smaller English force.

All the terrain maps in this book are drawn to the scale of 1in = 12in, so that a wargames table of 8ft × 5ft covers an area of 8in × 5in on the map. Keeping to that scale, draw a map 24in long × 15in wide to cover an area of 9 wargames tables, with the table-top battlefield taking up the middle oblong. In the surrounding oblongs continue such topographical features as hills, rivers, roads etc (see Fig 1). Movement on this map is at the same scale as movement on the wargames table, so that infantry who move 12in on the wargames table will move 1in on the map. To facilitate map movement, cover the map with a pattern of inch squares drawn with a mapping pen, or else with a transparent plastic sheet upon which similar squares have been marked. The practical use of such a map, combined with a Time Chart, is clearly demonstrated when setting up the Battle of Kadesh.

The wargamer commanding the force on the receiving end of

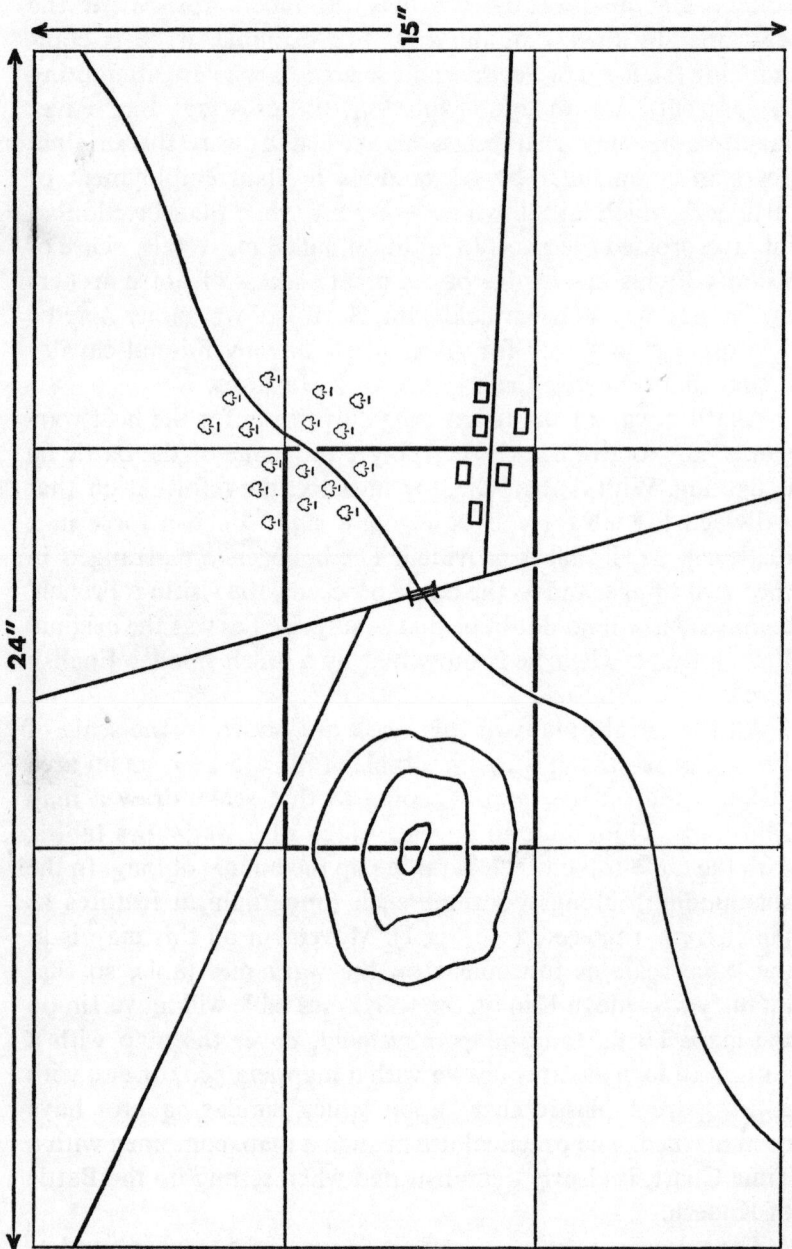

Fig 1

15"

24"

a surprise can have both the countryside of the battlefield (in the centre oblong) and its surroundings drawn inaccurately, so that it is difficult for him to assess the possible existence of an outflanking force, or the time of its arrival. The characteristic uncertainty of war is further demonstrated if the commander of the outflanking side is also provided with an inaccurate map, making his movements far harder than anticipated and giving the outflanked commander a slight chance of recovery. It is only fair to warn the wargamer that, when both sides have inaccurate maps, an umpire with an accurate chart will save a lot of argument!

A relatively simple method of simulating surprise and concealment is for both commanders to have their maps covered by a sheet of plastic upon which each move is marked with a chinagraph pencil. The umpire takes both plastic sheets and places them on top of each other, with his master map at the bottom, so that he can determine whether any troops are within visual distance of each other or have come into contact. This procedure is repeated at the conclusion of each game move, with the umpire checking both maps and handing them back without comment until a contact is made or forces become visible to each other. Then the troops can be placed on the battlefield as though they had come upon each other from behind concealing topographical features.

Terrain

The terrain over which the battle was fought played a material part in perhaps half the 15 battles: at Cynoscephalae the rough ground helped to break up the phalanx; at Taginae the charging Goths were channelled into a fire trap; at Bannockburn Bruce's carefully chosen position did much to win the battle for the Scots; at Morlaix and Auberoche both Northampton and Derby used woods to shelter their troops; at Neville's Cross the ravine on the Scots' right put them at a disadvantage; and at St Albans the undefended houses and gardens separating the King's forces lost him the battle.

Morale
Perhaps fortuitously, in none of the battles described in this
book did one force have a markedly higher state of morale than
its enemy. The only arguable exception is Bannockburn, for the
Scots' successes on the previous day had undoubtedly given
them a confidence that was raised even higher by the sight of the
disordered English at daylight. In all battles the morale of the
losing side falls as the battle progresses – Spartan consternation
at Leuctra when confronted by the unusual Theban tactics, and
dismay at Shrewsbury when Hotspur is killed. The 'morale'
section of most sets of rules make provision for the lowering of
morale of one side while raising that of the other. Since the
beginnings of recorded military history, a soldier's courage has
been raised or lowered by the same stimuli – fear of the unknown, death, mutilation, or disgrace – so that wargamers'
morale rules for all periods are very much alike. As in real life,
a smaller army can only defeat a larger on the wargames table
through advantages bestowed upon it by well-considered morale
rules. In this way, for example, the devastating effect of surprise
on the numerically superior French at Auberoche is simulated,
causing them to be defeated by a much smaller English force.

Weather
Not one of these battles was affected in any way by the weather.
There was no fog or extremes of heat or cold on any field,
although some of the losing commanders probably prayed for
the benison of a sudden concealing mist.

Finally, each battle can be magically transported through time
so as to provide a wargame in another era. How Craufurd's
Light Division of Peninsular War fame would have revelled in
the situation at Auberoche!

1

The Battle of Kadesh

1288 BC

IN 1292 BC Rameses II, an energetic young man of great ability, became Pharaoh of Egypt and set about restoring the Egyptian Empire to its former glories. His main enemies were the Hittites, who were strongly based in Syria, with Kadesh as the main bastion of their southern frontier. In the spring of 1288 BC Rameses took the field, and by mid-May was traversing the valley of the upper Orontes, a day's march from Kadesh in the plain below. In order of march the Egyptian army, 20,000 strong, was formed in 4 divisions:

The Division of Amon, with the Grand Guard led by the Pharaoh
The Division of Re
The Division of Ptah, including the Base Troops (recruits and young men)
The Division of Sutekh

When they reached the ford near Shabtuna, the troops halted to water and feed, and became strung out, causing a gap of from 1 to 2 miles between each division. Three miles from the ford the Forest of Baui slowed the column and allowed the division of Amon to get even farther ahead as the impatient Rameses, anxious to besiege Kadesh, marched on without waiting for the remainder of his army. By mid-afternoon, after covering 15 miles, Rameses reached a suitable camping area 1 mile west of Kadesh and some 8 miles north of the ford. Camp was pitched,

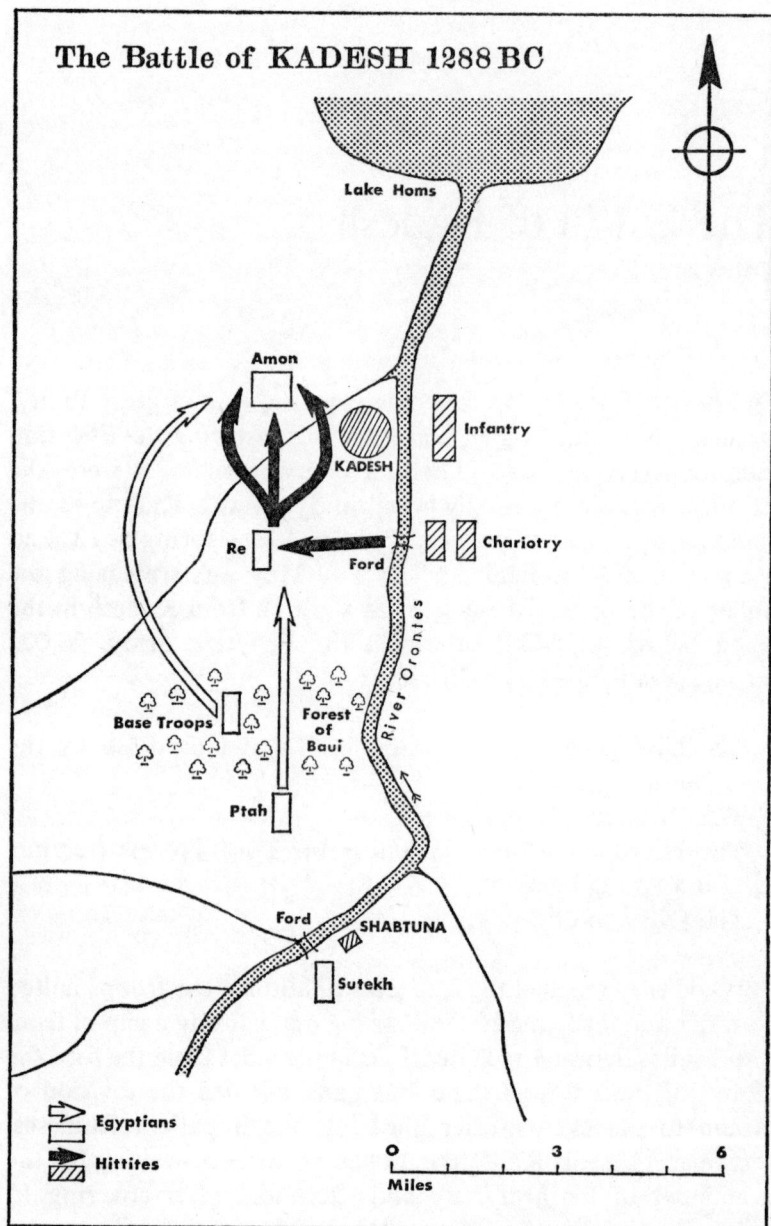

The Battle of KADESH 1288 BC

Lake Homs

Amon

KADESH

Infantry

Re

Chariotry

Ford

Base Troops

Forest
of
Baui

River Orontes

Ptah

Ford

SHABTUNA

Sutekh

Egyptians

Hittites

0   3   6
Miles

with a surrounding zariba of shields and Pharaoh's pavilion in the centre, chariots were parked and horses picketed.

Suddenly a rider galloped into the camp crying that a host of Hittite chariots had crossed the river to strike the divided Egyptian army, hitting the unguarded flank of the Division of Re and routing it. Mutallu, the Hittite commander, had completely outwitted Rameses with false information of his whereabouts, inducing the Egyptians to move along the left bank of the river while, on the right bank, the Hittite army kept the hilltop town of Kadesh between them and the enemy. In this way, completely unseen, they were able to close in from the south-east before fording the river 2 miles south of Kadesh to cut the Egyptian army in two and isolate Rameses, who was now opposed by vastly superior numbers.

He had only sufficient time to send his Vizier for help before fugitives from the Division of Re came pouring into the camp carrying men and tents before them. Then, as their pursuers came on the disordered scene, Rameses rallied his men and cut his way out to the north. Instead of pursuing, the Hittite charioteers began to plunder, as did the infantry formations following them up. Given vital time to regroup, Rameses flung his chariots in charge and countercharge on the weak enemy eastern flank by the river.

Some 4½ miles farther south the Vizier's chariot, keeping on the fringe of the battle, had come upon the detachment of Base Troops just emerging from the Forest of Baui; they were ordered to strike off to the left and, from the west, attack the Hittites in the captured camp. Continuing, the Vizier next met the Division of Ptah, which he led in a frontal attack from the south. Still south of the Orontes, the Division of Sutekh was too far away to intervene in the battle.

Arriving at the camp, the Vizier led forward the Na'Arun troops, a crack Canaanite unit, in phalanxes 10 ranks deep, to fall upon the now disorganised Hittite troops. Heavily engaged with Rameses to the north, assailed by the Base Troops from the west, and now attacked by these crack troops from the south, the Hittites looked around for support; but none was forthcoming, as the remaining 6,000 Hittite infantry stayed in its ranks on the far side of the river. Mutallu, having already committed all his chariots in the first attack on the centre, probably realised that his infantry were powerless against the Egyptian chariots. It is recorded that the Hittite King was drowned attempting to cross the river – Egyptian reliefs show him being pulled feet first out of the water.

The Hittite charioteers wavered, then turned and fled in a terrible rush for the ford, which was quickly blocked by a mass of struggling men and horses, and overturned chariots, into which the Egyptians poured a pitiless stream of arrows until there was not a single Hittite left on the west bank. The Hittite survivors retired into Kadesh to prepare for a siege, but the battered Egyptians also pulled back without attempting to take the city – so the battle ended indecisively.

Reconstructing the Battle

Although this battle was strung out over a front of about 2 miles, it has to be constricted to fit into a single wargames table, with the Hittite flank attack on the Division of Re taking place at one end and with the Amon camp at the other, allowing space for Rameses' counterattack and the final fighting when the Base Troops and the Divisions of Ptah arrive.

The organiser of the wargame should take the role of Mutallu, the Hittite leader, thus preserving surprise as the Egyptians lay out their forces with the Division of Amon in camp and the Division of Re approaching (see terrain map). Ideally, separate wargamers should command the Divisions of Amon and Re. 'Rameses' is not permitted to take any action until the frantic messenger arrives from the Division of Re; then he has to send out the Vizier, rally his forces and start counterattacking while awaiting assistance.

Certain problems of distance (ground scale) have to be considered: for instance, the 1½ miles that separated the Amon camp from the attacked Division of Re becomes 3ft on the wargames table, and the Vizier had to travel 4½ miles (9ft table distance) before reaching the detachment of Base Troops and a further $\frac{1}{2}$ mile (1ft table distance) to reach the Division of Ptah. Even if these forces react instantaneously, the Base Troops have to cover $4\frac{1}{2}$ miles (9ft) and the Division of Ptah 5 miles (or 10ft) before reaching the scene of action. These troops consisted mainly of charioteers, but the Wargames Research Group Ancient Rules (the most widely used rules for this period) allow chariots only an 8in move (18in when charging to contact), at which rate it will take the Vizier $13\frac{1}{2}$ game moves to reach the Base Troops and 15 game moves to reach the Division of Ptah. As these formations in their turn will take the same amount of time to reach the scene of action, Rameses' small force will have to face the might of the Hittite army for at least 27 game moves before the Base Troops arrive and 30 before the Division of Ptah comes to his aid!

As the majority of wargames are fairly conclusively settled in about a third of this time, the game is speeded up by introducing an illusory factor in the form of Chance Cards to determine the speed at which the Vizier and the reinforcing troops move. Chance Card No 1 allows the Vizier 4 game moves to reach the Base Troops and 5 to reach the Division of Ptah, and Chance Cards Nos 2 and 3 reduce his time further to 3 and 4, and 2 and 3, game moves respectively. A similar use of Chance Cards can then decide how long it takes the Base Troops and the Division of Ptah to reach the scene of action. A further set of these cards can be used to determine (a) whether the commanders of the Base Troops and the Division of Ptah are Above Average, so that they move quickly into action; (b) whether they are Average, taking perhaps a game move to rally their troops and set them in motion; or (c) whether they are Below Average, and fail to move at all. Of course, the last contingency will completely alter the trend of the battle, probably reversing the result. As we have said before, all movements should be recorded on a Time Sheet.

Commanders' Classification

Initially Rameses was outmanoeuvred, but he averted disaster by taking a firm grip on the battle. Mutallu, though beginning brilliantly, lost his advantage because he failed to exercise adequate personal command and control. Therefore, it would seem reasonable to class Rameses as an Above Average commander and Mutallu as Average. The Vizier, who played such an important part in averting an Egyptian defeat, was undoubtedly Above Average.

Number and Quality of Men
The Egyptian army are said to have been 20,000 strong, a large portion being chariots divided into groups of 50, and the rest infantry. The Hittite army totalled 17,000, half its strength being 3 men chariots containing a driver and 2 spearmen. Neither side used cavalry. The Hittite attacking force consisted of 2,500 chariots containing 7,500 men, and about 3,500 infantry. The remaining 6,000 Hittite infantry were never committed.

The wargamer may find such large numbers of chariots difficult to amass, so these forces may have to be considerably scaled down, or cavalry used instead of chariots. The main Hittite striking force should be just over twice as strong as the destroyed Division of Re, and about twice the size of the Division of Amon. The Base Troops (part of the Division of Ptah) could be about a fifth of its strength, and the whole Division of Ptah about the same size as the Division of Amon.

Once they had lost the advantage of their initial surprise, the Hittites suffered from a technical inferiority in weapons, since the Egyptian chariots, their crews armed with the long-range composite bow, were far more effective than the Hittite charioteers, armed only with spears and javelins. The Egyptians also possessed elite troops in the Na'Arun Canaanites. In the New Kingdom of Egypt (from 1580 BC onwards) charioteers and bowmen wore a helmet, and mail armour made of rectangular scales of metal. Infantrymen carried a small shield and used spears as their basic weapon. The composite bow, supplied to both charioteers and infantrymen, could penetrate the armour of the time at a reasonably short range.

Morale
At the outset the Hittite morale was first class, whereas that of their first opponents, the Division of Re, must have fallen to second or even third class by the surprise flank attack. As the battle progressed, the Hittite morale faltered, whereas that of the Egyptian Divisions of Amon and Ptah and the Base Troops remained consistently high.

Style of Fighting
Egyptian chariots operated as a unit, charging in a solid mass to drive the enemy before them, and manoeuvring so as never to allow themselves to be individually surrounded. The Hittite method of chariot attack was to charge the enemy pell-mell, and, if the fighting became too congested for manoeuvre, dismount the driver and turn him into an extra spearman. The spearmen on both sides formed in large groups, rather like a primitive phalanx, while the lightly armed archers and slingers swarmed out on the flanks and skirmished in front.

Terrain
The general terrain was most important, as it enabled the Hittites to surprise the Egyptians, whose formations, after pausing at the ford and traversing the Forest of Baui, became strung out and vulnerable; but all this happened beforehand, and does not need reproducing on the table. All that is required is a gently undulating area dotted with patches of scrub and perhaps small clumps of trees.

Military Possibilities
1   Rameses does not allow his columns to straggle, but waits at the ford of Shabtuna until each division has crossed, and then moves forward as a united force. This will probably save the Division of Re from surprise attack.
2   Though surprised, the Division of Re is not routed.
3   After routing Re and descending on the Amon camp, the Hittites press their advantage instead of halting to plunder.

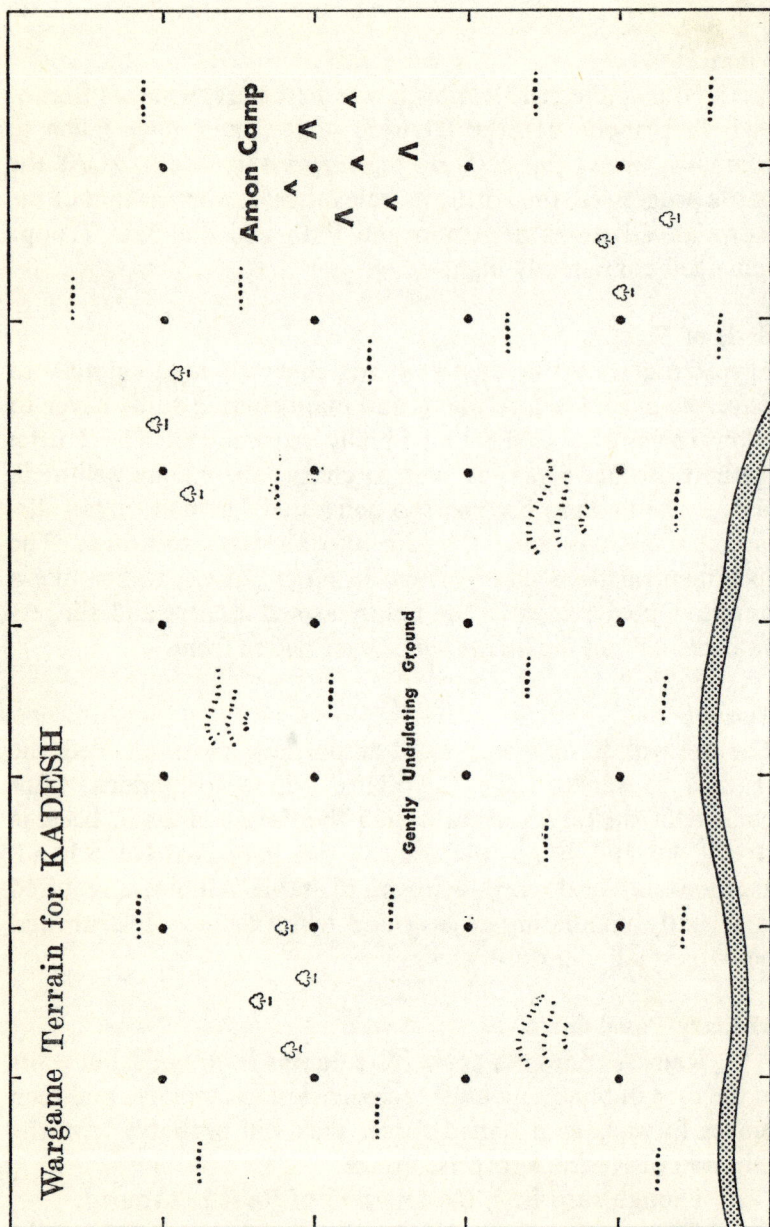

Wargame Terrain for KADESH

Amon Camp

Gently Undulating Ground

4 Rameses and his Division of Amon are completely dispersed, unable to rally and counterattack.

5 Either the Vizier is not sent or he does not reach the oncoming Egyptian columns.

6 The Base Troops and/or the Division of Ptah do not arrive in time to aid Rameses. The classification of their commanders can have a bearing on this.

7 Instead of remaining on the far side of the river, Mutallu, in the later stages of the battle, brings over the remaining 6,000 Hittite infantry; or he exploits to the full his initial success by immediately committing all his chariots, and despatching the mass of his infantry by the same ford or round to the north of the city to back them up.

In some ways Kadesh resembles Salamanca, with the Egyptians filling the French role but doing what the latter failed to do by turning defeat into victory.

2

The Battle of Leuctra
371 BC

IN ALMOST every battle the offensive takes one of 4 forms:

Flank attack – representing a massive human weapon aimed at the enemy's exposed side
Envelopment – like pincers squeezing from both sides
Penetration – a human wedge driven deep into the opposing force
Frontal attack – a relatively crude move, when not forced upon a commander by tactical limitations, generally indicating lack of skill

Attacks from the rear could be included, but they usually denote an ambush or a pursuit.

The first 3 methods of attack are closely related to each other in that they are basically forms of 'flank' attack, with 'envelopment' achieving its effect by hitting both flanks, and 'penetration', after rupturing the enemy's line, exploiting the 2 new flanks created.

Soldiers and armies not unnaturally feel distinctly uneasy when threatened from flank or rear, so that these primary tactical principles largely affect morale. Given sufficient time, defenders can face about to meet a single flank attack, but the threat to their morale is heightened by a compound assault on flank and front. Success for the troops making the main attack depends upon the ability of their comrades to 'fix' the other part of the enemy formation.

The Battle of LEUCTRA 371 BC

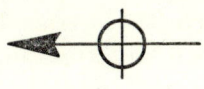

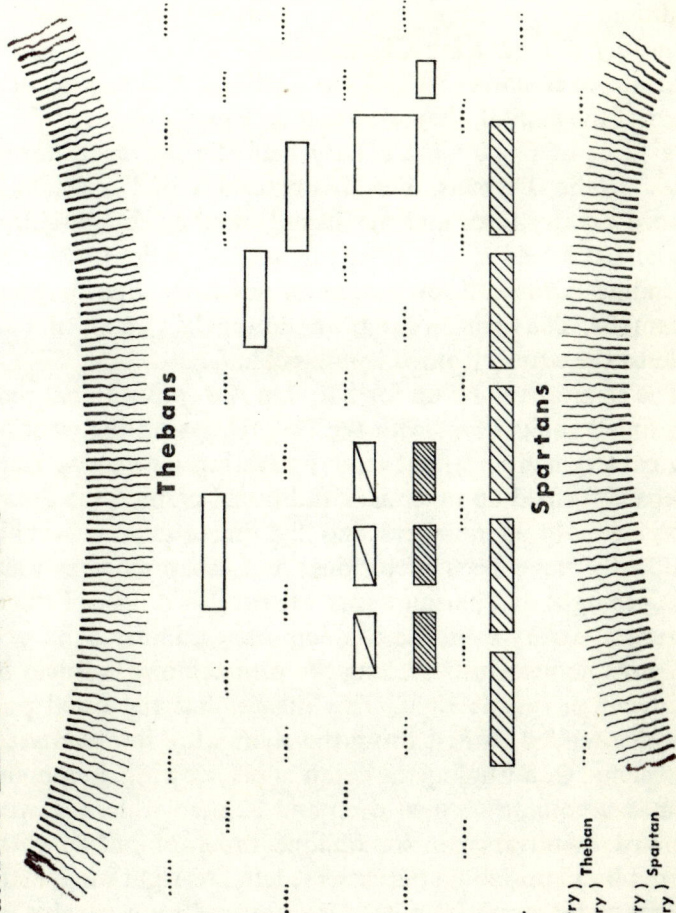

Thebans

Spartans

Infantry) Theban
Cavalry)

Infantry) Spartan
Cavalry)

Not to Scale

Military history presents many notable examples of flanking, but few are more remarkable than its first recorded instance in 371 BC, when Epaminondas at Leuctra inflicted an astonishing defeat on a hitherto invincible Spartan army, and killed their king, who had anticipated a simple frontal battle in the Greek tradition.

In July 371 BC King Cleombrutus I led an army of 10,000 Spartans to Boeotia to attack the Thebans. At Leuctra, 10 miles short of the capital, they were met by Epaminondas at the head of a force of 6,000 men. Nearly half of these were unreliable allies of the Thebans, but Epaminondas had a small, well-trained cavalry force and the Sacred Band of elite infantry.

The battlefield was a level unimpeded plain 1,000yd wide, extending between 2 low ridges on which the opposing armies encamped. The Spartan camp was entrenched on an eminence a little to the south, fronted by steep bluffs.

The Spartans drew up for battle in the conventional phalangial line 8 ranks deep, with the best troops on the right and a few cavalrymen and light troops covering the flanks. Had the Thebans formed up in similar traditional manner the Spartans, superior both in numbers and fighting qualities, would undoubtedly have been victorious; but Epaminondas was not going to fight on Spartan terms. He proposed instead to manufacture a victory from the contemporary military habit of each warrior unconsciously seeking the protection of a shield borne by the man on his right. This meant that the usual parallel advance tended to first bring the right wing into contact with the enemy. Quadrupling the depth of his left wing, Epaminondas formed a column 48 men deep and 32 men wide, thrown well forward in advance in an oblique order of battle, with the Sacred Band posted on the extreme left. His right wing, made up chiefly of his doubtful allies, was covered by a cavalry screen and echeloned to his right rear in thin lines facing the left and centre of the Spartan army. It was the first recorded example of the deep column of attack and a refused flank, forming a prototype for the holding operation combined with a main effort of a much later date.

Before the huge block of Thebans made contact, there was some sharp cavalry fighting on the Spartan centre/left, where the excellent Theban horse drove off the weak Spartan cavalry. At the same time the infantry on the refused right advanced slowly to occupy the attention of the Spartans to their front, but without actually engaging them. These unusual tactics hopelessly confused the Spartans.

Personally led by Epaminondas, the massive Theban column crashed home on the Spartan right to roll it up on its own centre by sheer weight of numbers. Then Epaminondas wheeled against the exposed flank of the remaining Spartans, so spreading the demoralisation of their shattered right wing to their centre and left, which had yet to strike a blow. At the same time the Theban centre and right made contact.

If the approach of the Spartan army had produced complete despondency in Thebes, no less did the threat to their rear convince the Spartans that they were hopelessly defeated, proving that flank attacks succeed in proportion to their menace to an opponent's rear. It was the conviction of disaster that caused the Spartans to flee, for at the moment of rout they still had more men on the field than the Thebans, while neither their left nor centre had been seriously engaged. The most successful outflanking operations in military history show the losers to have held the material advantage up to the point of panic, and their casualties to be largely the effect rather than the cause of their defeat. The fleeing Spartans left 2,000 dead on the field, including their king. Theban casualties were negligible.

At Leuctra Epaminondas introduced the concepts of concentration and economy of force coupled with a high standard of cavalry-infantry coordination. He gained a second triumph with his oblique order at Mantinea, a moral victory over the Spartans, although he was mortally wounded and, in consequence, the battle ended in less than a rout. Philip of Macedon was sent to Thebes as a hostage some 4 years after Leuctra, and undoubtedly learned much from the battle. At Issus in 333 BC his son Alexander defeated the Persians by adopting the tactics of Epaminondas.

C

Reconstructing the Battle

Perhaps this battle provides more of an interesting tactical exercise than a wargame, for both wargamers must conform to the tactical formations of both Spartans and Thebans. The events of the historical day should be followed in their correct sequence, though the 'Spartan' commander can enter into the spirit of the exercise by attempting to improve upon Cleombrutus' performance.

Commanders' Classification

Epaminondas completely outgeneralled King Cleombrutus. The Theban leader must obviously be Above Average and the Spartan Below Average.

Number of Men

The Spartans were formed into four brigades, totalling 2,100 men, and, with 6,000 allies and the complete Phocian levy, their force comprised 10,000 hoplites and 1,000 cavalry. The Theban force of 6,000 was composed of about 5,250 infantry, which included some 2,500 unreliable allies and the elite Sacred Band, plus about 750 first-class cavalry.

Quality of Troops

A warrior race, the Spartans must start this battle as first-class troops, although their cavalry seemingly were not quite so good and should be classified as Average. However, if the true spirit of the battle is to be caught and a well-deserved tactical victory gained by the numerically smaller Theban force, Spartan quality and fighting ability must fall after the initial successful Theban onslaught on their right wing. The Phocian levy could be classified as Average troops. Although it is recorded that about half the Theban infantry were unreliable allies, this fact is not reflected in the battle. Therefore, all the Thebans should be classified as first-class troops, with the Sacred Band and the cavalry force considered as exceptional or elite forces.

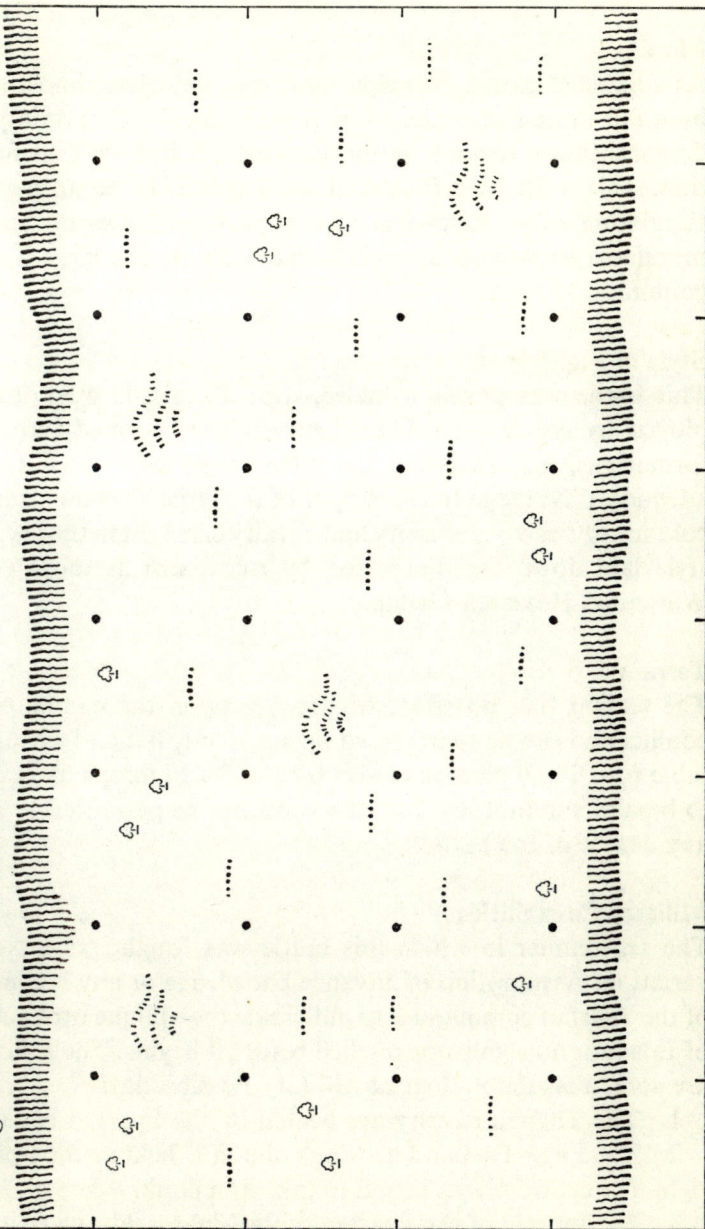
Wargame Terrain for LEUCTRA

Morale
As suggested earlier, the Spartans were defeated almost solely
by a fall in their morale, consequent upon the uncertainty and
demoralisation caused by the unusual Theban tactical formation. This must be reflected, if the battle is to be an accurate
simulation of its historical counterpart, by lowering Spartan
morale as soon as their right flank is hit by the large Theban
column.

Style of Fighting
This battle was purely a mêlée, with no missile weapons employed by either side. The spearmen were formed into close
formations, and rules for the mêlée must be slanted to give
adequate advantage to the weight of the large Theban attacking
column. Otherwise, infantry and cavalry can fight in the accepted
style laid down for the period by rules such as those of the
Wargames Research Group.

Terrain
The terrain had no effect whatsoever upon the course of the
conflict and can be represented by a perfectly flat and featureless
table top. Small clumps of shrub or rocky outcrops can be laid
to break its monotony, but they must not be permitted to affect
any aspect of the battle.

Military Possibilities
The set manner in which this battle was fought permits little
variation. Assumption of advance knowledge or any inclination
of the Spartan commander to anticipate the oblique order attack
of Epaminondas must be quelled before it begins. The following
are some reasonably logical Military Possibilities:

    1   The Theban cavalry are beaten by the Spartan cavalry.

    2   The large Theban left-wing column is held as the Spartan
right and centre pivots round to take it in flank.

    3   The morale of the Spartan centre/right holds, so that they
are not as readily dispersed as they were in fact.

3

The Crossing of the River Jaxartes

330 BC

THE CROSSING of the River Jaxartes was outstanding if only because no such advanced conception of artillery tactics was seen again until the last stages of the Roman Empire. Alexander and his Macedonians used cavalry as their main arm of battle, but relied greatly on their war engines, which, reduced to parts, could be carried on the soldiers' backs in areas inaccessible to pack animals.

In the 3 years since the Battle of Arbela, Alexander's campaigns had taken him to the Hindu Kush in modern Afghanistan, where his frostbitten soldiers laboriously made their way over a pass to descend on to desert steppes. At last they reached the River Jaxartes (also known as Syr Darya, the River of the Sands) where Cyrus the Great had been defeated and slain by the Scythians. On the far side of the muddy river, swollen by rains, appeared increasing numbers of these dreaded barbarians, who were armed with long swords and strangely curved bows, and whose long braided hair and baggy trousers made them one with their shaggy horses. They jeered at the Macedonians while keeping up a dropping fire of arrows that irritated Alexander's troops but caused few casualties.

The difficult river crossing had to be made, for to retreat south would invite the assembled Scythians to cross the river and harass Alexander's men unmercifully. For 3 days Macedonian engineers laboured to build 12,000 small wooden rafts and floats made from leather tents stuffed with hay. Others set up a series of gastraphetes (large mechanical crossbows), which

The Battle of the JAXARTES RIVER 330 BC

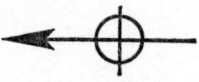

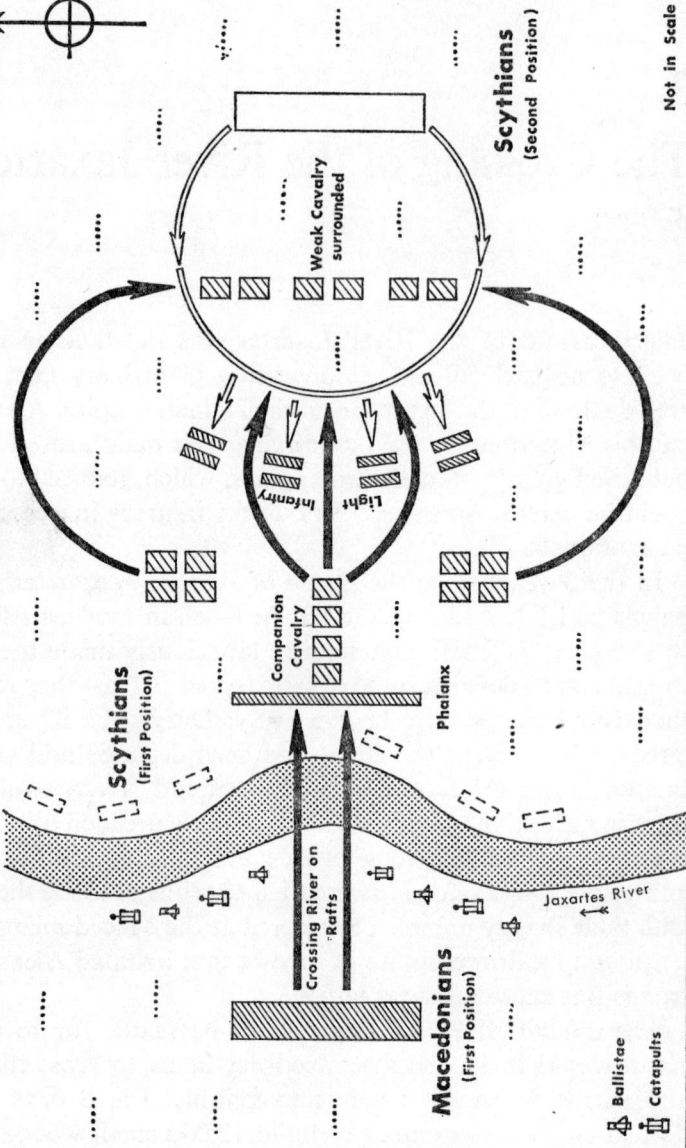

Scythians (Second Position)

Weak Cavalry surrounded

Scythians (First Position)

Light Infantry

Companion Cavalry

Phalanx

Crossing River on "Rafts"

Jaxartes River

Macedonians (First Position)

Ballistae

Catapults

Not in Scale

Alexander ordered not to be fired until the hour of the crossing. Then, with the rafts positioned in readiness and his troops lined up, Alexander gave the order to loose the barrage of crossbow javelins. The heavy missiles flashing across the water astonished the Scythians with their range and caused consternation by the ease with which they penetrated wicker shields and leather body armour. The barbarian horsemen promptly scattered away from the river bank and Alexander ordered the first rafts into the water. The Roman historian Arrian (Flavius Arrianus) later recorded the event:

> Seeing the Scythians confused by the discharge of his missiles, Alexander ordered trumpets to sound and keep on sounding. He led the first wave of the crossings. After the leading archers, javelin throwers and slingers had got to land, he ordered them to advance and harass the Scythians with missiles, to keep the horse archers from charging the first ranks of the infantry phalanx stepping out of the water, until all the cavalry had got across.

The enemy horsemen, braving the fire of the mechanical crossbows, tried to halt the crossing while the main Macedonian force was still on the water, but their firepower was matched by that of Alexander's light troops, and soon the Macedonian army was formed up in battle array on the north bank of the river.

Aware of the tactics of the Scythians and similar light cavalry nomads encountered in the Balkans and in Central Asia, Alexander knew that they had to be brought to battle and prevented from following their usual practice of circling groups at a respectable distance while harassing them with archery fire, and then overwhelming any small bunches separated from the main body. When attacked in force, these elusive horse archers could vanish like a mirage, releasing showers of arrows over their shoulders as they went.

Alexander, therefore, sent forward a greatly outnumbered cavalry force which, as he anticipated, soon came under fire from encircling horse archers. Next he sent out a skirmish line of about 2,000 light infantry (those who had served as a screen during the river crossing), which advanced in crescent formation with orders to engage but not to attack the Scythians who lay

between them and the surrounded Macedonian cavalry. Behind
the light infantry and hidden by them were 2 outlying cavalry
squadrons (hipparchies) and farther back in the centre, close to
the river bank and out of sight, were the Companion Cavalry.
Behind them the infantry phalanx formed up at the river's edge.

The fire from the light infantry screen diverted the Scythians'
attention from the Macedonian cavalry vanguard but they continued to circle that force while sending groups of mounted
archers to range up and down the slowly advancing Macedonian
skirmish line. They made numerous attempts to drive back the
light archers and javelin throwers, but found them to be as
elusive as they were themselves.

At the trumpet call the two flank columns of Macedonian
cavalry charged through their own skirmish line to trap several
thousand surprised Scythians in the area between the cavalry
vanguard and the light infantry screen; at the same time the
surrounded cavalry counter-charged to the rear. The central
column of the Companion hipparchy, closely followed by the
phalanx, then charged into the centre of the milling mass of
confused barbarian horsemen. Suddenly the Scythians, who a
few moments before had been part of an encircling force, found
themselves completely surrounded. It is recorded that of the
4,000 or so involved, at least 1,000 were killed, including a
known chief, and several hundred captured in the swirling dusty
battle that followed. Macedonian losses were greater than this,
but they had won a decisive moral victory.

Reconstructing the Battle
With the Scythians massed on their side of the river and the
Macedonian formations lined up waiting to cross, the battle
will begin when Alexander's 'artillery' opens fire. Local rules
must be devised so as to make their onslaught so devastating and
surprising that the Scythian cavalry inevitably withdraws from
the river bank. The 'artillery' will continue firing until it is
masked by its own infantry (the terrain could be so constructed
as to form a high bank that allows the artillery to fire over the
heads of the light infantry). If realism is required, the Mace-

donian light infantry can be ferried across the river on their rafts and floats, but it is really only necessary to move them across to the far bank, utilising a specified move-distance or part thereof for the operation. The fire-fights and mêlées will be fought in the usual manner, although, as far as the horse archers are concerned, it might be necessary to study those sections of the Wargames Research Group rules that cover evasion.

Commanders' Classification

Alexander, the Macedonian leader, had had some experience of dealing with elusive horse archers, and his tactics showed how well he had profited by that experience. He must be considered Above Average, and the outgeneralled commander of the Scythian cavalry Below Average.

Number and Quality of Men

It is estimated that there were about 4,000 Scythian light horsemen in this action, and that the Macedonian force consisted of about 2,000 light infantry, 1,000 heavy infantry (possibly a chiliarchia or taxis of 1,024 men ranged in a phalanx with a 64-man frontage), and 3 cavalry hipparchies (one of which was the famous Companions), each consisting of 512 horsemen.

The well-trained and disciplined Macedonians were all first-class troops, with the Companions forming an elite body. The use of covering fire by the gastraphetes was centuries in advance of its time. The Scythians, although irregulars, were natural warriors, and horse archers over the centuries (as at Carrhae in 53 BC, Pharaspa in 36 BC and even up to the Crusades of the twelfth century), were always difficult to handle.

Morale

Initially the morale of both sides was first class. Then that of the Scythians was dented by the surprise onslaught of Alexander's artillery, followed by the successful river crossing, so that thereon it must be considered slightly lower than that of the Macedonians.

Style of Fighting

The Macedonian peltasts and psiloi skirmished in the same

Wargame Terrain for JAXARTES RIVER

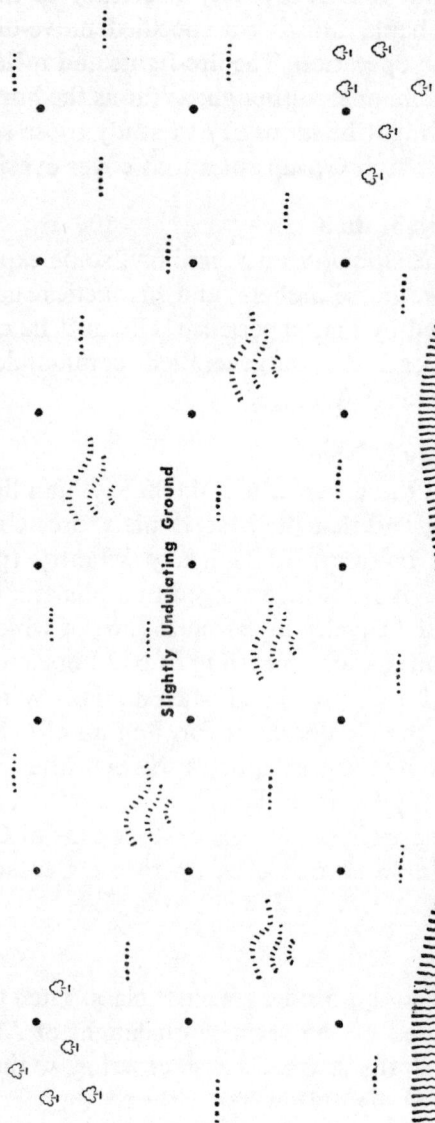

Slightly Undulating Ground

River Jaxartes

mobile fashion as Wellington's riflemen in the Peninsula, and Alexander's heavy infantry phalanx, besides being a base of manoeuvre for the shock action of cavalry, was a highly mobile formation capable of attacking at the run while retaining its formation. Cavalry was a decisive arm of the Macedonian army, and it was trained to combine with infantry in a manner superior to anything that had gone before. The Scythian horse archers, who displayed both mobility and firepower, gained their successes by galling and harassing slower moving infantry and cavalry with their showers of arrows until, in their attempt to pin down the elusive horse archers, bodies of men became separated from their main force, to be surrounded and wiped out at leisure.

Terrain
In this battle the river was no particular obstacle because of Alexander's effective method of crossing it. However, the final advance of the Macedonian cavalry, although screened by light infantry, may well have been aided by undulations in the ground.

Military Possibilities
This battle can be halted in its tracks if the Macedonians are wiped out in the river or slaughtered piecemeal as they come ashore. However, that would completely destroy the authenticity of the battle and is an eventuality that can be prevented by giving Alexander's artillery a high initial morale effect, so as to drive the Scythians back from the river bank.

1   The Macedonian cavalry decoy force is destroyed before Alexander can bring his other cavalry into action.

2   The Macedonian light infantry screen is caught, destroyed or scattered (a) on the river bank or (b) when moving forward to harass the Scythian horse archers surrounding the cavalry vanguard.

3   The Macedonian trap is not sprung completely, so that the Scythians avoid being hemmed in.

4
The Battle of the Ticinus River
218 BC

IN THE Po Valley, after the Carthaginian army had crossed the Alps, Hannibal and the Roman General Publius Scipio were eager for a confrontation, yet both were surprised to encounter each other so quickly.

At the head of 20,000 infantry and 2,000 cavalry, Scipio had crossed the River Po at Piacentia and advanced upstream on the north bank of the river while Hannibal, knowing Scipio to be in the area, was marching downstream on the same bank with the river on his right. At the Ticinus River Scipio marched his army across a bridge of boats to the western bank, where they camped for the night. Next day, leaving the legions in camp, Scipio took out a reconnaissance force of 2,000 cavalry and 4,000 velites (light infantry), marching through flat country screened by light scrub and trees. Towards the end of the afternoon they saw in the distance the dust clouds raised by the movement of a large enemy cavalry force. Scipio drew up in fighting formation, his light infantry in front flanked by Gallic horsemen, and the remainder of the cavalry in the rear.

Estimating the enemy force at about 6,000 cavalry, the Roman leader was in no hurry to commit himself to a fight, but so rapid was the Carthaginian advance that an engagement was forced upon him. Hannibal himself, leading the advance guard, quickly formed up his force so that his Carthaginian cavalry were in the centre and the light Numidian horsemen on the flanks.

Without pause both sides engaged, the cavorting and wheeling

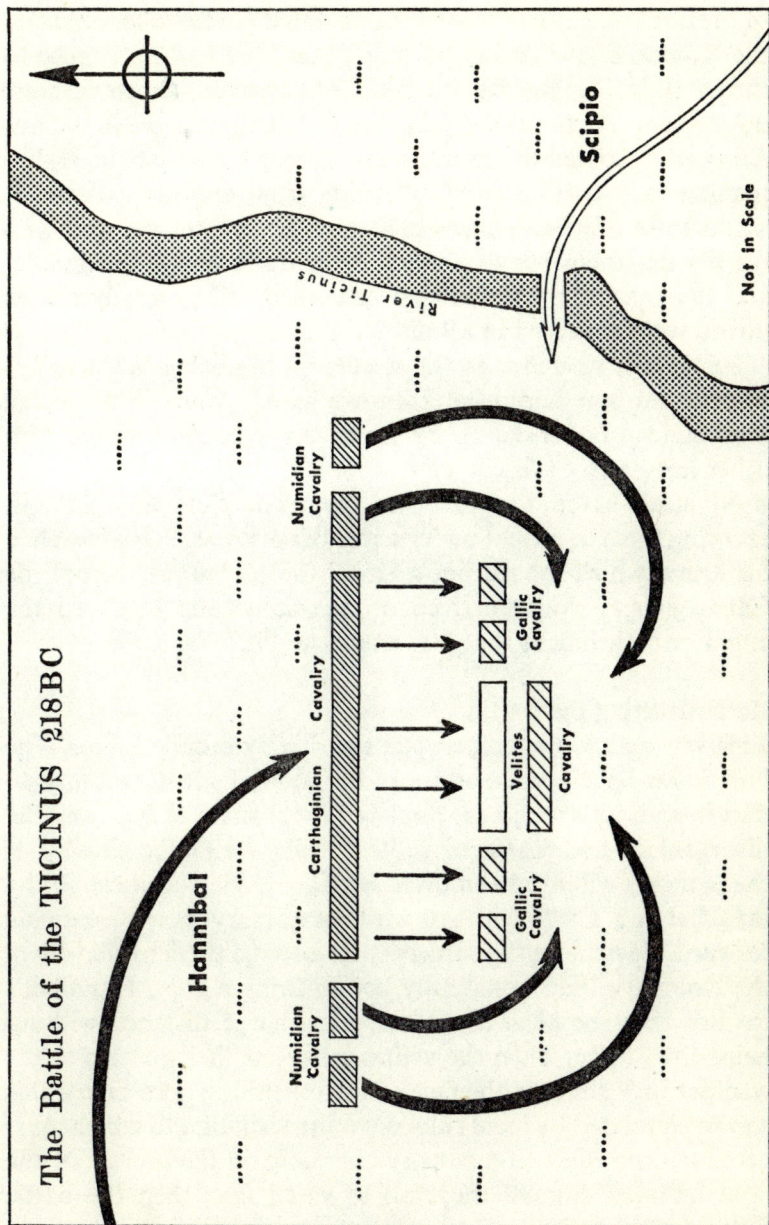

The Battle of the TICINUS 218BC

Hannibal

Numidian Cavalry

Carthaginian Cavalry

Numidian Cavalry

Gallic Cavalry

Velites

Cavalry

Gallic Cavalry

River Ticinus

Scipio

Not in Scale

of the horsemen raising clouds of dust that confused and blinded the Roman light infantry so much that they had no chance to throw their javelins before the Carthaginian horsemen were upon them. At the same time the fast light-armed Numidians fell on the Roman flanks and rear in their characteristic style – rushing in, retreating, and returning from another quarter. It was a form of attack that the Roman horsemen, engaged with the Carthaginian heavy cavalry, were quite unable to handle; and the light infantry, buffeted, trampled and generally ridden down, were scattered in all directions.

Scipio was wounded in the mêlée, his life being saved by his 17-year-old son Scipio (later known as Africanus) who, seeing his wounded father cut off by the enemy, charged forward with other horsemen to rescue him.

Nightfall saved the Romans from complete annihilation, allowing them to retreat under cover of darkness. Scipio reached his army, which moved back across the bridge just before the Carthaginian cavalry arrived on the scene and captured the engineer detachment that was destroying it.

Reconstructing the Battle
This was an encounter between two highly mobile forces. The Numidian light horsemen must be allowed great mobility to dart in and out around the flanks of the Roman cavalry and the disorganised and scattered velites, while the Gallic cavalry of the Romans will fight at a disadvantage, having been caught by the charging Carthaginians while stationary or just moving forward. To achieve the surprise that allowed them to ride down the Roman velites, apparently before firing a shot, Hannibal's cavalry must be allowed to come to charge distance without being fired upon. Then the velites must test their morale to see whether they stand in the face of the onrushing horsemen – this can be simulated by local rules covering visibility, to which must be added the effect of a cavalry charge upon the morale of the light infantry. Should they fail to stand fast, then the battle follows its historical course.

Commanders' Classification

Seemingly neither commander had much effect upon the course of the battle once it was joined. Hannibal may have ordered the Numidian flank attacks, but they are more likely to have been their normal battle tactics. Considering that Scipio was wounded during the course of the action and recalling Hannibal's fame as a military leader, it is justifiable to class Hannibal as Above Average and Scipio as Average.

Number and Quality of Men

Scipio's Roman force consisted of about 1,000 Gallic horsemen, 1,000 Roman cavalry, and 4,000 light infantry. Hannibal led some 6,000 cavalry – about 3,000 Numidian light horsemen and the same number of Carthaginian heavy cavalry.

The organisation and discipline of the Gallic auxiliary cavalry of the Roman army was less formal and rigid than that of the legion, while the Roman velites, lightly armed with javelins and darts, were the youngest, least-experienced soldiers, whose agility and lack of military formalism made them suitable for skirmishing in front and on the flanks. This rather unfortunate combination of forces was quite unable to cope with the Carthaginian heavy cavalry and the Numidian horsemen, who, like the Scythian horse archers, were natural warriors. Besides being better handled, the Carthaginian cavalry was superior to both the Roman and Gallic horsemen, who had no answer either to the swift attacks and retreats of the Numidians. As their light infantry were not allowed even to join in the battle, the Roman force was immediately reduced to only a third of the enemy's numbers.

Morale

Initially both forces were probably equally confident, although Hannibal's reputation and his march over the Alps may have shaken the Roman morale. However, from the onset of the swift Carthaginian attack the Roman morale dropped considerably. Rules usually allow for states of morale to fluctuate

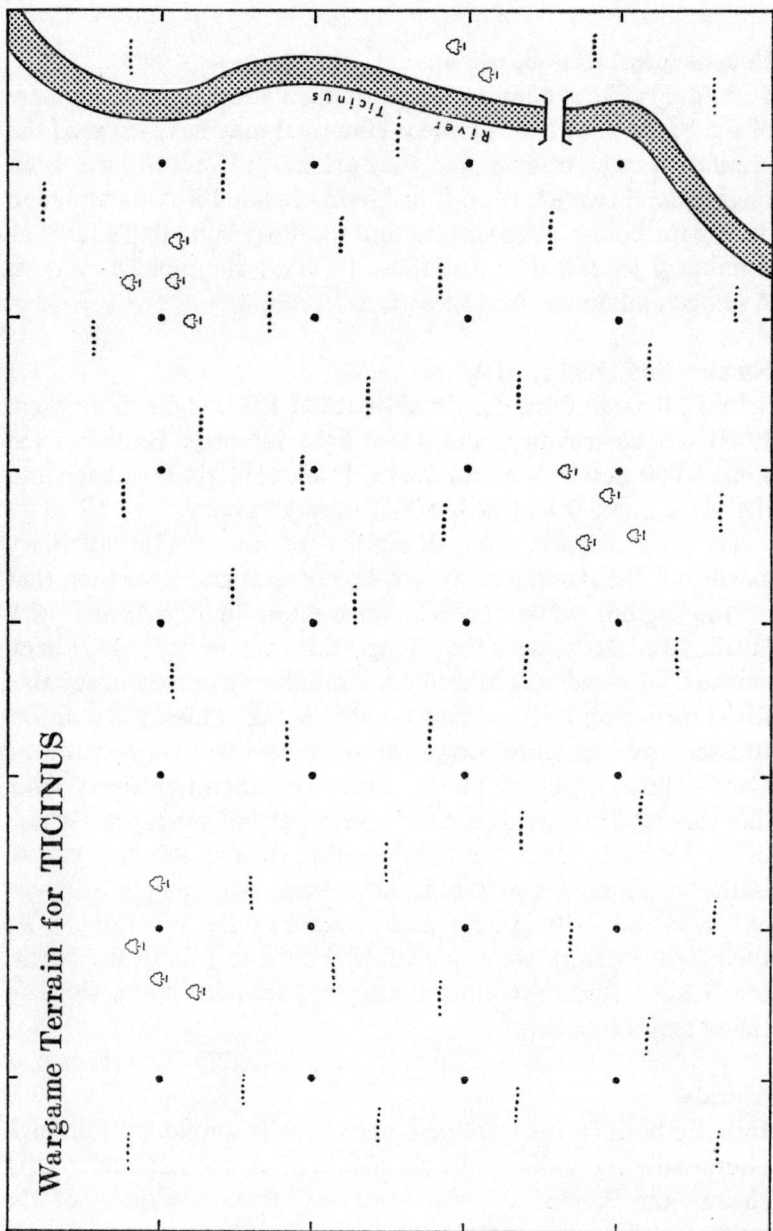

Wargame Terrain for TICINUS

River Ticinus

in accordance with the course of the battle, and this aspect must be emphasised in this battle if accuracy is to be maintained.

Terrain

Except for the fact that the River Ticinus probably prevented the Romans from refusing an engagement, even if they had wished, the terrain played no part in this action. Presumably it was a flat, featureless and dusty area dotted by light scrub and small trees, highly suitable for cavalry manoeuvring.

Military Possibilities

1   The Carthaginian cavalry does not attack so quickly.
2   The velites were able to join in the battle.
3   The Roman cavalry puts up a better fight.
4   The Numidian flank attack is held, presumably by the velites (as at the River Jaxartes).

Foreign auxiliaries were fighting for both sides – Gallic cavalry for the Romans, and Numidian horsemen for the Carthaginians. Probably the Carthagian heavy cavalry were also mercenaries. There might be the Military Possibility of unreliability about foreign allies or mercenaries, a factor that could be introduced into this action or associated with morale.

D

5

The Battle of Cynoscephalae
197 BC

FOUGHT 5 years after the great battle of Zama and the first occasion on which the flexible Roman legion and the Macedonian phalanx met in the open, the affair at Cynoscephalae was the Jena of Macedon, revealing to the incredulous Greeks that the phalanx had met its master, that the descendants of Alexander's soldiers had given way at the first shock to the 'unknown quantity' of the Roman army. This supreme test of the merits of the rival formations occurred during the Second Macedonian War (200–196 BC), fought in Greece and Asia. It was a conflict in which the successors of Alexander of Macedon took on the increasingly powerful Romans, who were allied to most of the Greek cities. Arriving at Heracles in early April 197 BC, Titus Quinctius Flaminius linked up with the Aetolian army of 6,000 foot and 400 horse under Phaeneus, and with Amynander and 1,200 Athamanians, bringing his army up to about 26,000 men, including 2,400 cavalry. Flaminius led them into Thessaly as Philip of Macedon moved towards him from Larissa at the head of an army of some 23,500 foot (including 18,000 Macedonians) and 2,000 horsemen, which had been quickly enlisted and trained, and included even boys of 16.

There was some desultory fighting near Pherae before both armies turned west seeking better ground, and lost touch with each other, although they were marching on parallel lines – Philip to the north and Flaminius to the south of the Cynoscephalae hills (Kardagh). In these hills near Scotussa, on a misty morning, covering detachments of both armies clashed in

The Battle of CYNOSCEPHALAE 197 BC

Macedonian Camp

Army in Line

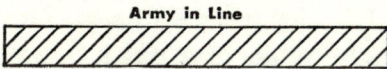

Column

Elephants

Right Wing

20 Maniples

Phalanx

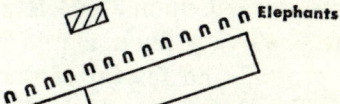

Left Wing

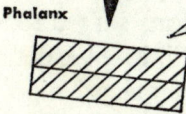

Army in Line

Roman Camp

Hilly, rocky Ground

Not to Scale

a small action that grew larger as reinforcements came up, until both sides were heavily engaged. The Romans were pressed back, despite some bold charges by the Aetolian horse, as Flaminius deployed his army facing the hills, before advancing his left to meet the enemy. Encouraged, the Macedonians moved forward to occupy the southern slopes of Kardagh, although the unexpected general engagement was not entirely to Philip's liking, as the Macedonian left wing had yet to come into position and, more important, the ground over which they were to fight was broken and unfavourable to the phalanx.

The battle fell into three separate and successive actions. First, on the west, Philip descended from the hills with the right half of the phalanx to drive back in great disorder the Roman left, personally commanded by Flaminius. At this reverse the Roman commander rode across to his hitherto inactive right wing and ordered the legions and their allies, preceded by some elephants, to fall upon the left half of the phalanx, at that moment deploying from march column. Disordered by the broken ground and fearful of the elephants, the Macedonian wing in this second part of the battle could not cope with the attack of the legions and the Aetolian infantry, and was broken and routed. The Roman right-wing formations surged forward in pursuit, and a Tribune, acting on his own initiative, detached 20 maniples, totalling 2,000 men formed of the principes and triarii, and led them to the left. Outflanking the victorious Macedonian right-wing phalanx, the maniples in this third and final period were able to attack the formation from the rear, breaking it and driving it from the field in confusion.

Realising that all was lost, Philip rallied the survivors and left the field, having lost about 13,000 men against Roman casualties of a few hundred.

Reconstructing the Battle
The battle can either begin with small parties of light troops and cavalry skirmishing in the centre of the table as they endeavour to win superior positions for their advancing armies, or with Flaminius deploying his army and advancing his left as Philip's

right-wing phalanx moves forward to attack him. This first phase of the battle is followed by the Roman and allied infantry on the right, supported by elephants, catching the Macedonian phalanx before it has formed. The third phase begins with the Tribune leading his 20 maniples against the Macedonian right-wing phalanx, which has already defeated the facing Roman formation. Too much time must not be allowed to elapse here, as no self-respecting wargamer is going to allow a victorious formation to stand and gloat while its comrades are being beaten.

Commanders' Classification

Neither Philip nor Flaminius showed up particularly well: the former accepted battle knowing the terrain to be unsuitable to his formations; and the latter was beaten on the left, though he deserves credit for sending his right wing into action after this defeat. Both may be rated as Average, but the unknown Tribune who led 20 maniples into the rear of the Macedonian phalanx should be classified as Above Average.

Quality of Men and Style of Fighting

The luck of the Romans at Cynoscephalae in encountering a hastily trained force operating on unsuitable terrain encouraged the legions to find their feet at an early formative period. There is no gainsaying, however, that the relatively fast and mobile legion possessed a tactical flexibility denied to the solid mass of the now slow and unwieldy phalanx, particularly in this battle when the latter was formed of ill-trained and inexperienced troops operating on rocky ground. Like the British square, the strongest feature of the phalanx was its cohesion; if thrown into disorder by the terrain or caught in the process of forming up, it was lost. In favourable conditions it pushed forward like an irresistible militant hedgehog to carry all before it.

The superiority of the Roman military system was emphasised a generation later at Pydna, when the legion again beat the phalanx, this time on level ground well suited to phalangial tactics. This defeat of Perseus (Philip's son) has been accepted by many military writers as conclusive of the merits of the two formations. Yet the time was past for the trial to be a fair one, as the phalanx was no longer at its best, partly because the length of the sarissa had been increased to 21ft, so limiting the manoeuvrability of the phalanx, as the men had to bunch tightly together for mutual support, grasping their huge spears with both hands. Even more incapable than before of changing front, the formation immediately broke up if assaulted in flank or rear. The contemporary practice of relying on unaided infantry to protect the flanks was not as successful as Alexander's use of cavalry for this purpose. The phalanx could could still perform wonders on level ground against an adversary who awaited attack, but when opposed to the mobile and flexible legion, resembled a bull assailed by nimble matadors. This tactical flexibility had been given to the legion by Scipio Africanus, and the troops employed by Flaminius were largely Scipio's veterans from Spain and Africa, where doubtless the Tribune who turned the battle had learned his tactical lessons. It was on the foundations laid by a more gifted commander, therefore, that Flaminius gained his victory at Cynoscephalae.

It can be claimed that the maniples led forward by the un-named Tribune played the biggest part in the victory. The 20 maniples (a maniple was composed of 2 centuries each of 60 to 80 men, with the triarii maniple consisting of 1 century only) were formed of principes or veterans, the experienced backbone of the army (armed with 7ft javelins and a broad-bladed short sword or gladius), and the older triarii. The javelins were thrown at the enemy just before contact, and then the principes closed with the sword in a tactical concept comparable to a bayonet attack preceded by rifle fire. The triarii were the oldest group of legionaries who, forming every third line of infantry, brought steadiness and experience to the formation. They were armed with a 12ft pike and the gladius.

The Aetolian allies of the Romans are frequently mentioned in accounts of the battle and must have been first-class troops.

The number of elephants is not known, but undoubtedly they played a big part in scattering the forming Macedonian phalanx.

Elephants possessed both enormous potentialities and great limitations. Horses often refused to face them, but disciplined formations could either turn them back or allow them to run through intervals in their ranks. They were most successful when used against troops to whom they were unfamiliar, for they could be stampeded by disciplined and resourceful opponents to become more dangerous to friend than foe.

Morale

Both forces could start the battle with the same standard of morale, or the Macedonians, a relatively scratch force, might start with a lower morale than the Romans, but the morale of each in turn must be affected by the fluctuating progress of the battle. Morale rules could be slanted to allow the Macedonians to fight well until the moments of crisis, when they disintegrate rapidly. These moments are when the left wing encounters the elephants, and when the victorious right wing is hit in the rear by the maniples, and both circumstances requiring strong and immediate morale-reaction tests.

Terrain

All important in this engagement, the terrain was sharply undulating and extremely rocky, as might be expected on the slopes of a range of hills. It is probably best simulated on the wargames table by numerous low piles of books, pieces of wood or plastic ceiling tiles covered by a blanket or cloth. The surface should be liberally spread with pebbles and pieces of stone, scrub and small stunted trees to simulate the uneven ground that helped to destroy the phalanx. While your bemused opponent is still studying this tricky terrain, quickly assume the mantle of the Roman commander!

Military Possibilities

Of course, the prime possibility would be for Philip not to accept battle on unsuitable terrain, but that would mean no wargame. Other possibilities are:

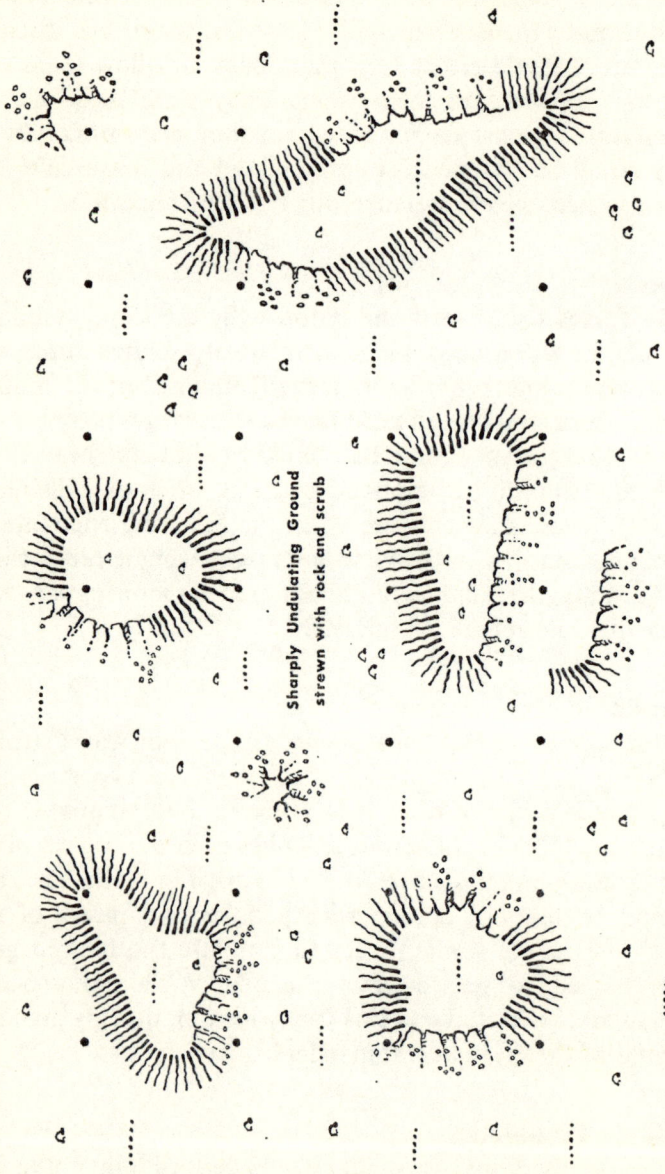

Wargame Terrain for CYNOSCEPHALAE

Sharply Undulating Ground strewn with rock and scrub

1   The victorious Macedonian right phalanx has the time or experience to exploit its success.

2   The Macedonian left-hand phalanx forms more quickly, and is not caught in relative disorder by the legions and the elephants.

3   The Tribune is not so alert, and does not lead his 20 maniples to attack the Macedonian right phalanx.

6

The Battle of Taginae

AD 552

THIS BATTLE bears a remarkable similarity to Crécy (1346), although Edward III had never heard of Taginae. In a sense it reflects the decline of Western military art, for 8 centuries passed before similar successful tactics were evolved.

The battle took place near the Flaminian Way at the Apennine village of Taginae (Tadinum), sometimes known as Gualdo Tadino, in Umbria. In AD 552 Justinian I, the Byzantine Emperor, sent Narses with an army of some 20,000 men round the head of the Adriatic into Northern Italy. Crossing the Apennines, Narses encountered the Gothic leader and Christian King Totila (Baduila) with an army of about 15,000 men in a narrow mountain valley that could not be bypassed. The Goths were formed up in 2 lines, with heavy cavalry lancers in the front and archers and infantry armed with spears and axes in the rear.

Narses, an experienced commander, immediately drew up his army in a formation enabling it to employ the customary Byzantine tactics of hitting the enemy by a combination of firepower and shock – first wearing them down by archery and then subjecting them to successive charges of armoured horsemen. Aware of Gothic contempt for Byzantine infantry, who would seldom face a cavalry charge, Narses, gambling on the Gothic commander assuming them to be the usual despised mail-clad infantry, placed his dismounted Lombard and Heruli mercenary lancers in his centre, forming them up into a phalanx of spearmen. On each of their flanks he placed 4,000 Roman foot

58

The Battle of TAGINAE AD 552

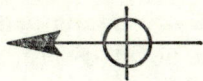

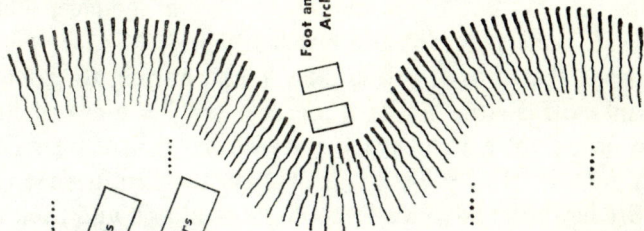

Foot and Horse Archers

Cataphracts

Foot Archers

Dismounted Lancers

Byzantines

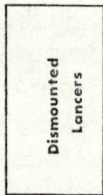

Cataphracts

Foot Archers

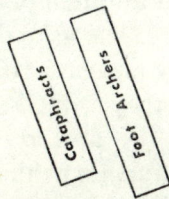

Goths

Heavy Cavalry Lancers

Heavy Infantry

Archers

Archers

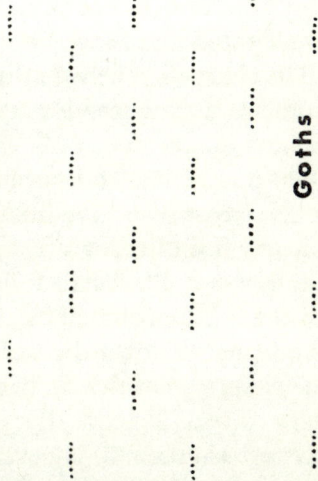

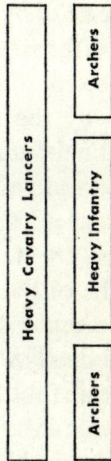

Not to Scale

archers, angled forward in a crescent, like a wide-mouthed box. Behind the archers were the heavy Roman cavalry cataphracts. Well thrust out to the left on a dominant height, not near enough to invite attack but close enough to be a menace, Narses stationed 2 bodies of foot and horse archers.

At about the time of the midday meal, hoping to catch the Byzantines off guard, Totila flung forward his mass of heavy cavalry, attempting to crush the Byzantine centre by striking directly into it. Sweeping up the valley, the Goths first came under fire from the advance force on Narses' left, and then were raked with flight after flight of arrows from the 8,000 archers of Narses' left and right wings. The stricken Gothic horsemen were thrown into confusion by the uncontrollable kicking and plunging of wounded and riderless horses, and were left leaderless by the premeditated shooting down of their distinctively clad chieftains. The charge slackened into a trot and then into a walk until all momentum had been lost, so that the cavalry were powerless to break into the fence of outstretched spears that faced them. While this action had been taking place, the rear ranks of the Gothic cavalry and the infantry that followed had been under the heavy fire of arrows from the flanks, where the mobile Byzantine horse archers had foiled all attempts of the Gothic foot bowmen to support their cavalry. The second line of Gothic infantry advancing down the valley had naturally been left behind by the charging cavalry, so that they formed an ideal massed secondary target for the Byzantine archers, whose showers of arrows caused them to lag even further behind.

Towards eventide the Goth leader Totila was wounded, at which the greatly reduced and disheartened Gothic cavalry wavered before wheeling about to ride blindly back up the valley. At this, demoralisation swept through the surviving infantry, whose only contribution to the battle had been to act as targets for the Byzantine archers.

Seeing the Goths falling back, Narses ordered his spearmen and archers to open their ranks, and the heavy cataphracts came sweeping through the gaps to ride down the fleeing cavalry and scattered infantry. The Goths were caught in a double envelopment by the Byzantine cavalry, and more than 6,000 of them, including their leader Totila, were killed.

The Battle of Taginae was one of the first-recorded successes in the coordinated employment of pike and bow, and was made possible by the Byzantines' balanced combination of archers, infantry and cavalry.

Reconstructing the Battle
This is largely a matter of laying out the Byzantine army in its historical formation and awaiting the Gothic attack, which should be in the recorded 2 lines of cavalry and infantry. Using standard methods of casualty assessment and morale state, it is difficult to see how the battle can go any other way than it did in AD 552. It is a fascinating exercise and example, however, of the highly advanced (for its day) Byzantine concept of firepower and shock.

Commanders' Classification
Narses was undoubtedly an Above Average commander. The best that can be said for Totila was that he was an Average leader, fighting in the customary Gothic style.

Number and Quality of Men, and Style of Fighting
Narses commanded an army of about 20,000, composed of mercenary lancers, Roman cataphracts, and Roman mounted and dismounted archers. The Gothic force of 15,000 men was formed of armoured lancers riding heavy horses, heavy infantry and light archers.

The Byzantine archers were well-trained and disciplined soldiers; the mercenary lancers were first-class professionals; and the cataphracts were the most reliable soldiers of the Dark Ages and the mainstay of the Byzantine army, combining firepower, discipline, mobility and the ability to act as shock troops. The Gothic heavy cavalry lancers were capable of thundering charges that infantry were unable to withstand, and were reasonably disciplined. It is possible that they had some of the characteristics of the feudal knights of a later date. The Goths'

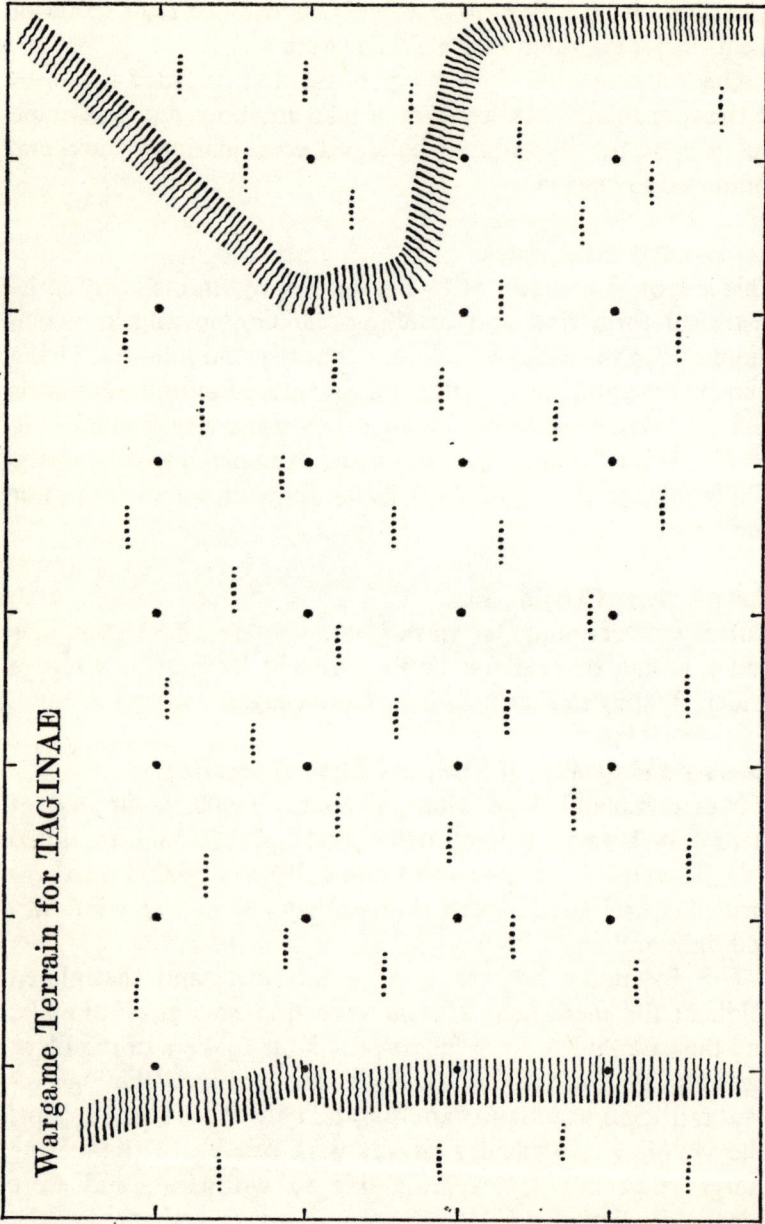

Wargame Terrain for TAGINAE

heavy infantry and archers were sturdy and reliable soldiers.

Morale
Undoubtedly both sides were full of confidence at the start of the battle, but morale must have fluctuated as it progressed. Thus, the evident success of Byzantine firepower must have increased their confidence while steadily lowering the morale of the Goths. These factors are automatically reproduced by standard morale rules.

Terrain
Apart from the advantageous position taken up by the Byzantine horse archers on the eminence forward of their left, the terrain is only notable because it channelled the Goths into a fire-trap, although the battlefield was initially of their own choosing.

Military Possibilities
1 The Goths attack a flanking group instead of the Byzantine centre.
2 The Gothic cavalry breaks the Byzantine centre.
3 The Gothic infantry manages to come into action.
4 The Byzantine detachment thrust forward on their left is under poor control, and is ineffective.
5 The seemingly shattered Goths manage to rally in the later stages of the battle and withstand or rout the cataphracts.

7

The Battle of Stamford Bridge
1066

IN SEPTEMBER 1066 Harald Hardrada, King of Norway, and the English King Harold's renegade brother Tostig landed their Norse army from a fleet of more than 300 vessels and ravaged the north-east of England. They crushed the East Riding Fyrd (local levies) under Earls Edwin and Morkere at Fulford.

In London King Harold Godwinsson assembled his personal bodyguard – the Housecarls and the Thegns – raised the Fyrd, and set off for the North, gathering reinforcements en route. Although only a small proportion of the army was mounted, the Saxons marched over 180 miles in little more than a week, to arrive in Yorkshire without the invaders being aware of their coming.

On Monday 25 September, about 5,000 men (two-thirds of the Norse army) marched 14 miles from their base at Riccall to Stamford Bridge for a council with local leaders. It was a hot day and the Norsemen, burdened with helmets, shields, spears, swords and axes, cast aside their byrnies, the knee-length coats of leather sewn with iron rings and studs that formed the standard armour of the age throughout Western Europe.

The wooden bridge at Stamford crossed the River Derwent at a point where it broadened suddenly into shallows in a valley approximately ¼ mile wide. On either side of the reeded and sluggish stream the land rose some 50ft, on the east to a plateau later to be known as Battle Flats. Arriving in the peaceful area at midday, the Norsemen threw themselves down on the grass or wandered by the river. Discipline was relaxed and scouts

The Battle of STAMFORD BRIDGE 1066

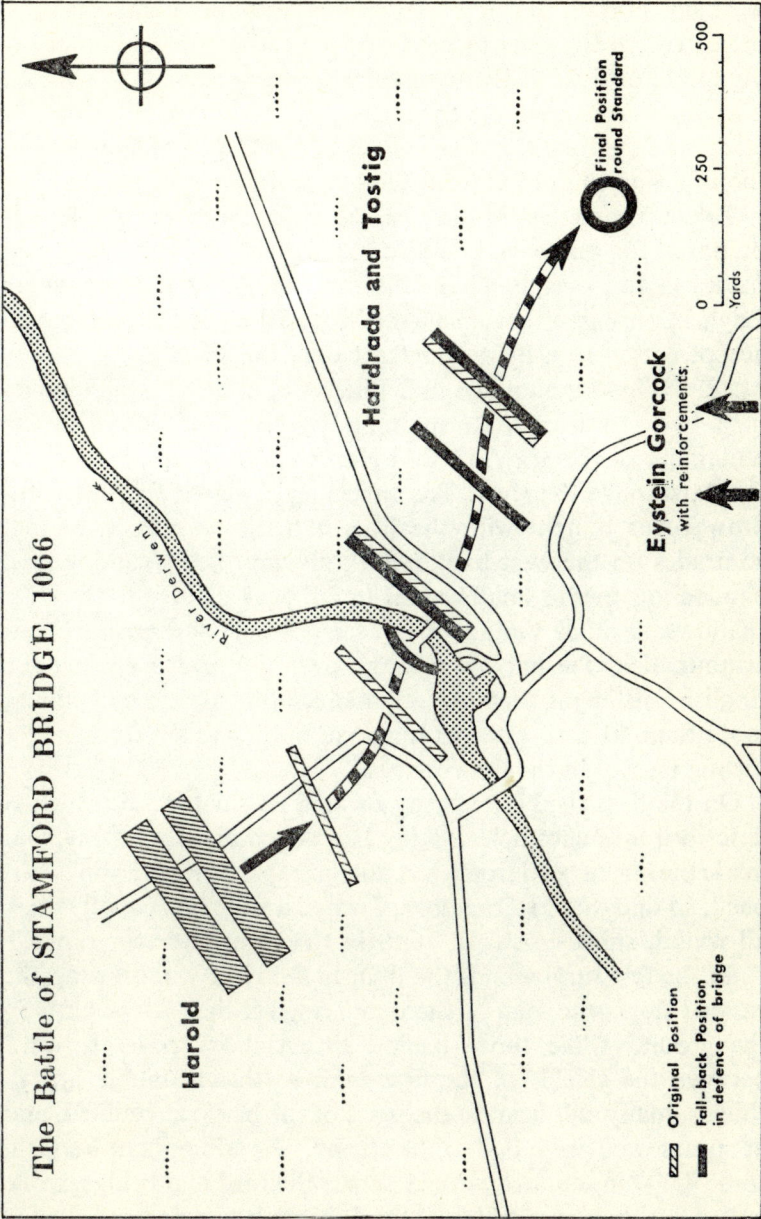

Harold

Hardrada and Tostig

River Derwent

Eystein Gorcock
with reinforcements;

Final Position
round Standard

0   250   500
Yards

▨▨▨ Original Position

███ Fall-back Position
in defence of bridge

E

were not posted. Leaving behind their accoutrements, some of them crossed the river to herd grazing cattle. Suddenly a growing cloud of dust to the west thickened and spread as men on horseback appeared on the ridge, their banners bearing the devices of the English King – the Golden Dragon of Wessex and the Fighting Man of Harold Godwinsson.

Taken completely by surprise, the Norsemen were wholly unprepared for battle, only 100 or so wearing body armour and those on the west bank, in the direct path of the advancing English, only lightly weaponed. They had no alternative but to accept battle, as a fighting retreat over the 14 miles to Riccall presented insurmountable difficulties. So, after sending 3 horsemen back to Riccall to summon Eystein Gorcock with the remainder of the army, Harald Hardrada drew up his force on the flats above the river. The time they required for this withdrawal was bought with the lives of their scantily armoured comrades on the west bank who, knowing they were doomed, banded together in small groups to fall back fighting to the river shallows. A single Viking, one of the few who had retained mail armour, held the narrow passage over the bridge against the English onslaught, beating off attack after attack and killing more than 40 men with his long axe until stabbed from below through a gap in the floorboards.

On the flats the Norse army formed its shield wall, which is described by a chronicler: 'King Harald arranged his army, and made the line of battle long, but not deep. He bent both wings of it back, so that they met together; formed a wide ring equally thick all round, shield-to-shield, both in the front and rear ranks.'

As the few survivors of the fight in the river withdrew up the eastern slope, the men of the outer ring set their spear butts in the ground while those behind thrust their spears forward between the shields of the first rank so that a double line of shields, many still bearing the scars of the battle at Fulford, and of spears were presented to the enemy. The triumphant English, some 4,000 in number, surged across the ford and bridge, to be halted by their marshals on the left bank and formed up for attack.

Battle accounts vary. Writing in the thirteenth century, the Icelandic chronicler Snorri Sturlasn tells of the English cavalry attacking, supported by their archers, in a manner resembling that of the Normans at Hastings in the following month. Many historians, including Sir Charles Oman in *The Art of War in the Middle Ages* (1924), deny that at this date the English ever fought on horseback. They probably rode to battle and then fought on foot. It seems unlikely that Harold and his men would have fought in 2 such vastly differing styles at Stamford Bridge and Hastings within a few weeks of each other, and it is likely that they fought on foot at Stamford Bridge.

During a short truce Harold offered his brother Tostig generous terms but only 'seven feet of English ground or as much more as he may be taller than other men' to Harald Hardrada. Then the battle began with a clamour of warcries. Shouting 'God Almighty!' and 'Holy Cross!' the Housecarls and Thegns, followed by the mass of the Fyrd, advanced up the gentle slope. In the fierce fighting the Norsemen, although un-protected by mail armour, were a match for the axe-wielding Housecarls, and the shield wall held firm. But occasionally 'berserk' fury seized a Norseman so that he forgot discipline and prudence, breaking out of the shield wall and surging wildly down the slope at the English. The gaps in the shield wall were hastily filled as men closed in to each other, although here and there small groups of Norsemen fought desperately in isolation.

At this stage Harald Hardrada led forward his best men, the reserve, to gain time for the shield wall to reform. Fighting recklessly to clear a space around him, the gigantic figure pre-sented a perfect target for the Fyrd's archers, who brought him down with arrows through cheek and throat.

The hard fighting continued into mid-afternoon, with the ranks of the Norsemen thinning rapidly and the shield wall being crowded closely by the English. Both armies had lost heavily, but the English were getting on top – and there was no sign of Eystein Gorcock with the reinforcements from Riccall. At last the shield wall broke and the battle split into scattered knots of desperately struggling men. It was now that Tostig, who had taken up the Norse banner, was killed by an arrow.

Then, too late to affect the issue, Eystein, with the sons of Thorburg and Nicholas, together with the Orkney Earls and their men came running across the southern approaches to the flats. Surprised by these fresh foes, the English were close to rout, but recovered themselves, and the most bitter struggle of the day began. It did not last long, for the newcomers, wearing full armour, had hurried 14 miles from Riccall through the heat of the afternoon and were near exhaustion when they reached the battlefield. Eystein broke through to the Norse standard, but their attack lost its momentum and wavered as, urged on by the Housecarls, the Fyrd rallied. As the afternoon merged into evening, Eystein was killed, and the Norse resistance ceased as the survivors fled.

King Harold Godwinsson held the field, but he had lost many of the Housecarls and Thegns who might have brought him victory at Hastings a month later. Less than a twelfth of the Norse army survived the battle, the survivors being allowed to depart after swearing never to return.

Reconstructing the Battle
This engagement will take careful planning, for a smaller force, partly of untrained militia, must defeat a numerically stronger force of first-class veteran fighters. To simulate their historic victory, the same conditions as applied in 1066 must be reproduced: (a) about a quarter of the Norse force must be scattered on the west bank, with only a minimum of arms and no mail armour; and (b), as many as 75 per cent of the remaining Norsemen must fight without mail armour. The outcome of the battle was much influenced by the Norsemen's lack of armour. Some distinction must be made during the wargame between those who are fully equipped and those who are not, and this can be represented by 'Saving Throws' allowing a casualty not wearing armour to carry on fighting only if he achieves a dice score of 4, 5 or 6.

The disposition of the Norse army must reproduce its initial dispersal on both sides of the river. Penalties in some form or other will simulate the lack of scouts, and the fighting qualities of those Norsemen on the west bank, as affected by relaxation of discipline. The dispersed Norsemen delay the oncoming English to enable the rest of their comrades to form up and rally on the flats as Hardrada sends messengers to Riccall for reinforcements. These riders should be subject to the vagaries of Chance Cards. A strict check should be kept on the length of time taken by the English to overcome the Norsemen on the west bank, and this should be related to the time it takes the remainder of the Norsemen to rally on the plateau and form their shield wall. The 'berserk' fury that caused some Norsemen to break from the shield wall can be accommodated by local rules. ('Berserkers on the Wargames Table' are dealt with in D. F. Featherstone's *Wargames Campaigns* and *Wargames through the Ages*, Vol I.) The sole Viking defending the bridge can be the subject of a colourful 'skirmish'-style wargame.

The progress of the Viking reinforcements from Riccall can be controlled by Chance Cards. On arrival, they will fight for one game move as Above Average troops, but then their exhaustion will reduce them to Average.

Commanders' Classification

The speed with which Harold Godwinsson reacted to the invasion of his kingdom, assembled his force and sped northwards to surprise the invaders, coupled with the generally courageous bearing of his troops throughout the battle, indicates that he is an Above Average commander. Similarly, there is little doubt that Harald Hardrada had the most inspiring effect upon his men during their unexpected battle, and so he will share this rating. His second-in-command, Harald's brother Tostig, does not figure very prominently in the account of the battle and it is fitting to make him an Average commander.

Number and Quality of Men, and Style of Fighting

Some 5,000 men, two-thirds of the Norse army, were at Stamford Bridge, the remainder of their force remaining behind with the boats at Riccall. Harold Godwinsson's force, composed of his elite Housecarls, Thegns and the Fyrd, numbered some 4,000 men.

The Norsemen carried shields and wore helmets and byrnies, and they were armed with spears, swords and axes. Ferocious warriors, they were perhaps the finest fighters of their day, who mostly fought to the death. The terror they aroused could invoke a 'ferocity' factor (see D. F. Featherstone's *Wargames through the Ages*, Vol I).

In the English army only the Housecarls were capable of matching the Norsemen's fighting ability, although, as was shown at Hastings a few weeks later, the English Thegns and the Fyrd were also capable of some good solid fighting. The House-carls were armed in the same way as the Norsemen and, as they showed at Hastings, were particularly adept in their use of the murderous two-handed axe. They were unlikely to go berserk. The Thegns, usually men of wealth and standing with some military training, were well armed and equipped with long straight swords, spears and light lances. The Fyrd consisted of peasants armed with swords, spears, javelins, clubs and primitive weapons made from agricultural implements; without protective clothing, they not only lacked military training but were almost completely undisciplined. There were probably 400 archers with the English force, forming part of the Fyrd, and their lower morale and fighting ability would have been compensated for by their ability to stand off in safety and fire at an enemy without the means to reply.

During the prolonged mêlée which forms much of this battle, all the Norsemen will count as Above Average fighters, as will the Housecarls, who formed perhaps a third of the English force. Half the Thegns will be considered Above Average and the other half Average fighting men, while half the Fyrd will be classified as Average and half Below Average.

Morale
It was second nature for the Norsemen to believe themselves to be better fighters than anyone else, and early success during this invasion had confirmed that opinion. The English, inspired by

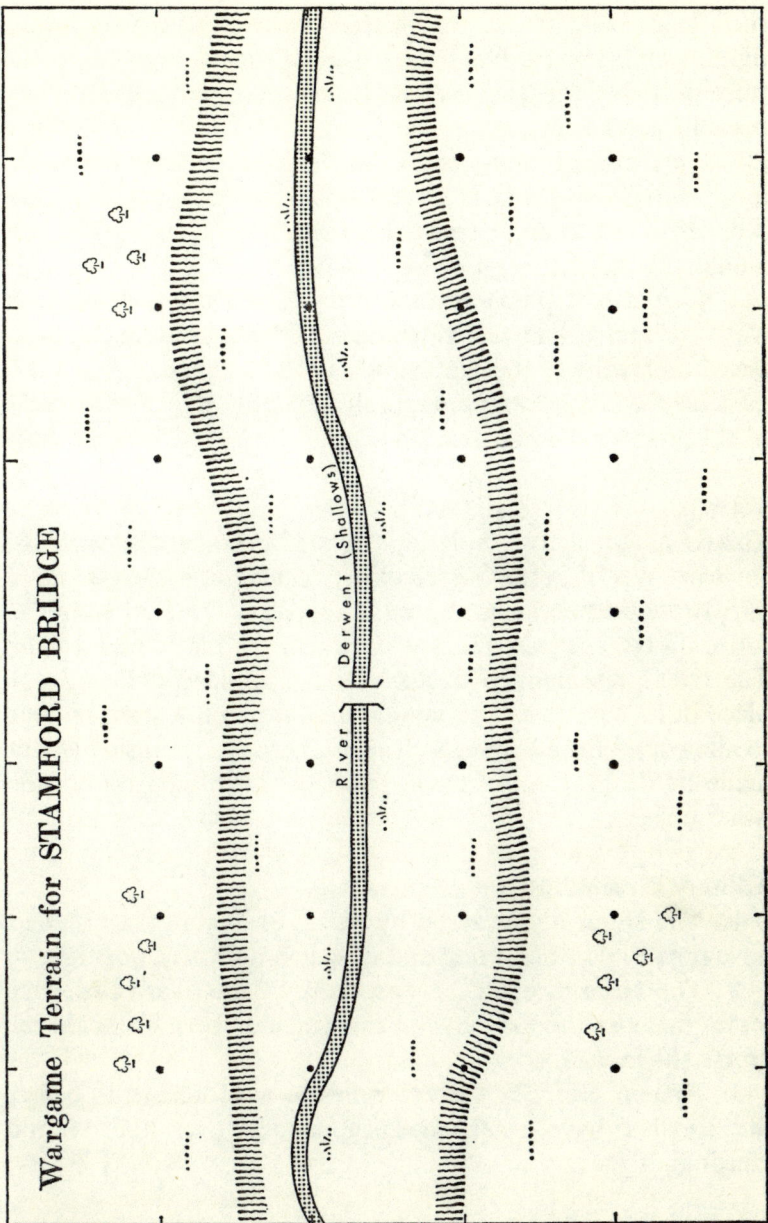

Wargame Terrain for STAMFORD BRIDGE

River Derwent (Shallows)

their leader and possessing in the Housecarls men of similar
fighting ability to the Norsemen, were of equally high morale, as
were probably the Thegns. The Fyrd would undoubtedly have
been in good heart, otherwise its members would have taken
advantage of the innumerable opportunities for desertion on the
long journey north. It is doubtful whether the morale of the
Norsemen fell to any appreciable extent during the battle, nor
would that of the English, once they realised they had caught
the Norsemen unprepared and that the battle was going their
way. Norsemen had a high standard of discipline and morale
based on loyalty to their immediate chieftain, however, and it is
possible that their morale might drop temporarily on the death
of their leader, Harald Hardrada.

Terrain
This is an attractive table-top terrain to construct, with the
shallow river meandering across its centre and crossed by a
narrow wooden bridge, and banks rising in gradual slopes to
form, on the east, the plateau on which the battle was fought.
The river presented no obstacle and prolonged fighting took
place in its shallows. The undulations were only steep enough
to allow minimal advantages. This is another terrain that can be
made by draping a cloth over strategically placed books and
other objects.

Military Possibilities
   1   The Norsemen, warned by clouds of dust in the distance,
have time to don their mail and rally in a defensive position.
   2   The Norsemen on the west bank of the river cause such
losses that the English have insufficient numbers to tackle the
Norse shield wall.
   3   Eystein Gorcock arrives with the reinforcements before
the English have established supremacy over the Harald
Hardrada's men.

8

The Battle of Lewes
1264

In MAY 1264, with a small army of about 4,500 infantry and 500 cavalry, Simon de Montfort marched from London into Sussex to meet the numerically superior army of King Henry III and his son Edward (later Edward I, the 'Hammer of the Scots'). Based on Lewes, that part of the Royal army commanded by the King camped south of the town around the Cluniac Priory of St Pancras, and Prince Edward's force was based on Lewes Castle, which flew the flag of its governor, de Warenne.

Wearing a white cross on back and breast, de Montfort's men left their camp before daybreak on 14 May 1264, marching behind the crest of a ridge towards Lewes. They were seeking an advantageous strategical position from which to attack, for it would have been fatal to draw up in the regular battle array of the day and await the onslaught of the King's superior force. When they eventually formed their line of battle at the crest of the ridge, a force of 500 Londoners under Nicholas Segrave and Henry de Hastings was placed on the left, while Henry de Montfort and his brother Guy, Simon's sons, commanded the right, the Earl of Gloucester led the centre, and a body of elite cavalry under Simon de Montfort himself was held in reserve. The baggage, left behind the ridge, included a waggon mounting de Montfort's standard and certain prisoners or hostages.

De Montfort's army advanced unseen down the slope before clashing with a foraging party whose survivors carried the alarm to the town. By then de Montfort was hardly more than a mile from the Castle and no more than $1\frac{1}{2}$ miles from the Priory.

73

The Battle of LEWES 1264

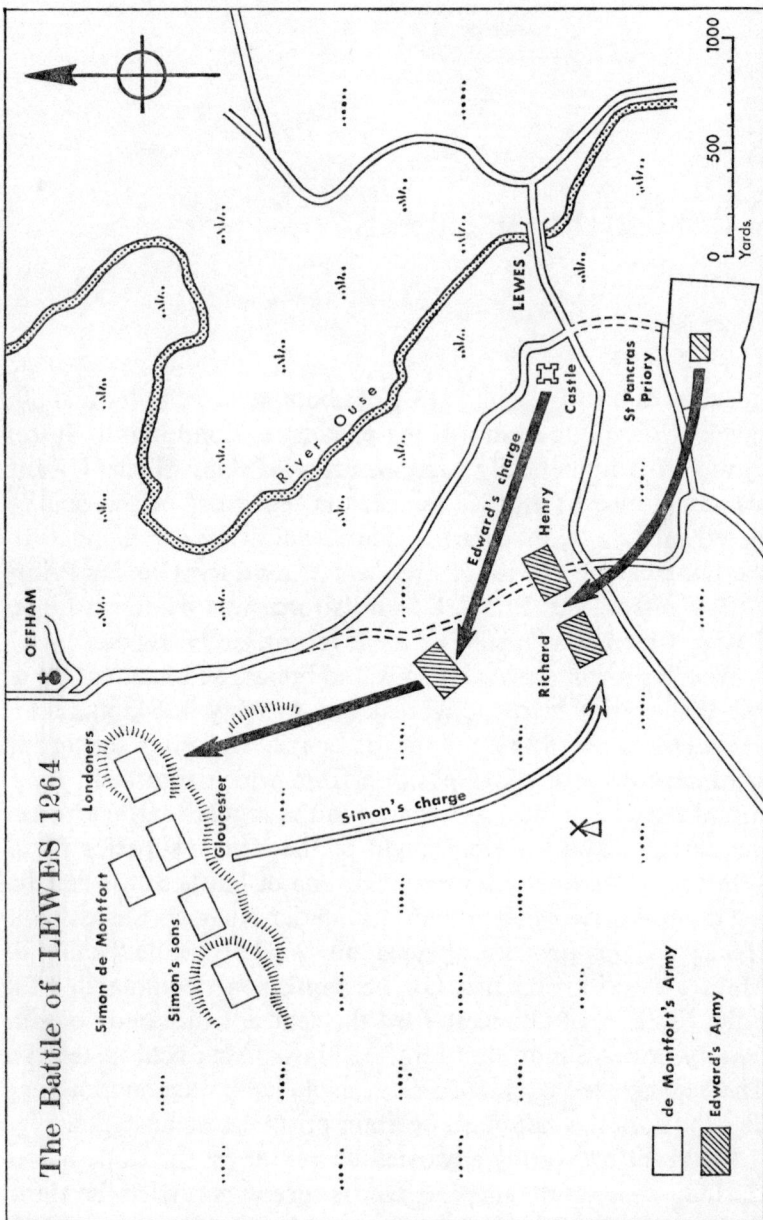

OFFHAM

Londoners

Simon de Montfort

Gloucester

Simon's sons

Simon's charge

River Ouse

LEWES

Castle

Edward's charge

Henry

Richard

St Pancras Priory

Yards.

0   500   1000

de Montfort's Army

Edward's Army

Prince Edward and his men armed themselves hastily and galloped from the Castle, to find themselves confronting the Londoners, who were untrained levies. It has been suggested that they had been given the task of firing the town and delivering a surprise attack from the rear once contact had been made elsewhere along the line. Inevitably, as soon as Edward's cavalry charged them, they turned and fled, to be pursued with great slaughter for some 3–4 miles.

Meanwhile, after much confusion and many counter-orders, the King's men had moved out of the town and taken up a position on some high ground. Seeing Edward's cavalry disappear into the distance, Simon de Montfort realised that this left the Royal infantry at the mercy of his own cavalry, and that, if he could force the King to surrender before the Prince returned, the day would be won. He advanced his centre $\frac{1}{2}$ mile to engage Henry's infantry, and a confused mêlée took place as Gloucester came to close quarters with the centre of the Royal army, led by King Henry in person. A contemporary account records that 'the King was much beaten with swords and maces, two horses were killed under him and he escaped with difficulty'. At the same time de Montfort's sons led the right wing on the easiest line down the slope to fall on the left wing of the Royalists, commanded by Prince Richard, whose men fought stubbornly and well until they were forced from the field. Richard, after taking refuge in a mill, was compelled to surrender.

Then de Montfort sent his mounted reserve round his right flank to attack the King's left flank. The shock of 500 horsemen galloping downhill was not to be resisted, and Henry's left broke completely beneath the avalanche. De Montfort could then concentrate his whole strength on the Royal centre, where King Henry was fighting valiantly, surrounded by a group of faithful knights and men-at-arms. The battle raged furiously around the Royal Standard until the Royalists were broken. As the Royalists scattered, trying to escape, many rushed towards the river, to flounder in the treacherous mire of the marshy lands around it.

Prince Edward and his knights came back from their disastrous pursuit as dusk fell, halting to plunder Simon's baggage, and slaughtering the unfortunate prisoners. Then, finding the King defeated and the battlefield controlled by de Montfort's men, the Prince, unable to reach the Castle over which the flag of de Warenne still flew, rode to the Priory, where he and his father both sought sanctuary until, later that night, they surrendered.

Reconstructing the Battle

This is another action in which a smaller force defeats a considerably larger one – not particularly easy to reproduce on the wargames table. To balance the disparity in numbers and to simulate the effects of the poor light of early morning, de Montfort's advance can be controlled by Chance Cards, which will also determine at what stage the Royal army becomes aware that it is under attack. Prince Edward's men must be first in the field and must encounter the 500 London infantrymen, whose rout is inevitable. The subsequent pursuit takes place under the terms of the Feudal Cavalry rules (see *Wargames through the Ages*, Vol I, 20–21).

With the Royal cavalry out of the way, de Montfort's infantry, through its surprise attack, is able to contain the Royal infantry, though outnumbered 6 to 4. So the battle progresses until de Montfort's 500 horsemen are able to bring it to a conclusion. Prince Edward and his cavalry have about as much impact upon the battle on their return to the field as Prince Rupert's men did on rejoining Charles I at Edge Hill nearly 400 years later.

Commanders' Classification

The battle was won because de Montfort's skill in handling his men made up for his inferiority in numbers. Probably he purposely sacrificed his London levies to Prince Edward and his cavalry, knowing that the impetuous young Prince would pursue them from the battlefield. Simon de Montfort is definitely an Above Average commander. King Henry's valiant conduct in the mêlée does not signify extraordinary leadership, and both he and his son Prince Edward, although the latter was to become one of England's greatest warrior kings, must be classified as Below Average.

Number and Quality of Men

De Montfort commanded an army of about 4,500 infantry and 500 cavalry to oppose King Henry's force of about 6,000 infantry, with 1,000 cavalry under his son Prince Edward.

Both armies were typically feudal, their mounted knights and men-at-arms the 'kings of the battlefield', although likely to take off in a wild and formless pursuit at any moment. Undoubtedly de Montfort's cavalry reserve was exceptionally well handled for the day and age. Both in prestige and effectiveness the infantryman of the period was decidedly inferior to the cavalryman, and the most that was hoped of him was that he would stand firm while the cavalry got on with winning the battle. De Montfort's foot soldiers engaged in the usual medieval 'sprawling brawl' with Henry's infantry until such time as de Montfort's cavalry crashed home and ended the battle. It is rather surprising that there were apparently no archers present on either side.

Morale

It is likely that de Montfort's men possessed a high standard of morale because (a) although knowing themselves to be inferior in number, they purposefully took the offensive; (b) they had great confidence in their leader; and (c) de Montfort himself, inspired by the recently ended Crusades, had turned his cause into a religious one, so that the white crosses his men wore carried more significance than identification.

Before the battle began it is quite probable that the Royal army, conscious of its numbers and with the King as its commander, had an equally high morale, but the early morning surprise attack obviously shook it. Therefore, the King's infantry should be classified as inferior to their attackers in morale, especially when they are charged by cavalry and their own cavalry is nowhere to be seen. Prince Edward's cavalry, fighting their own useless part of the battle, displayed the

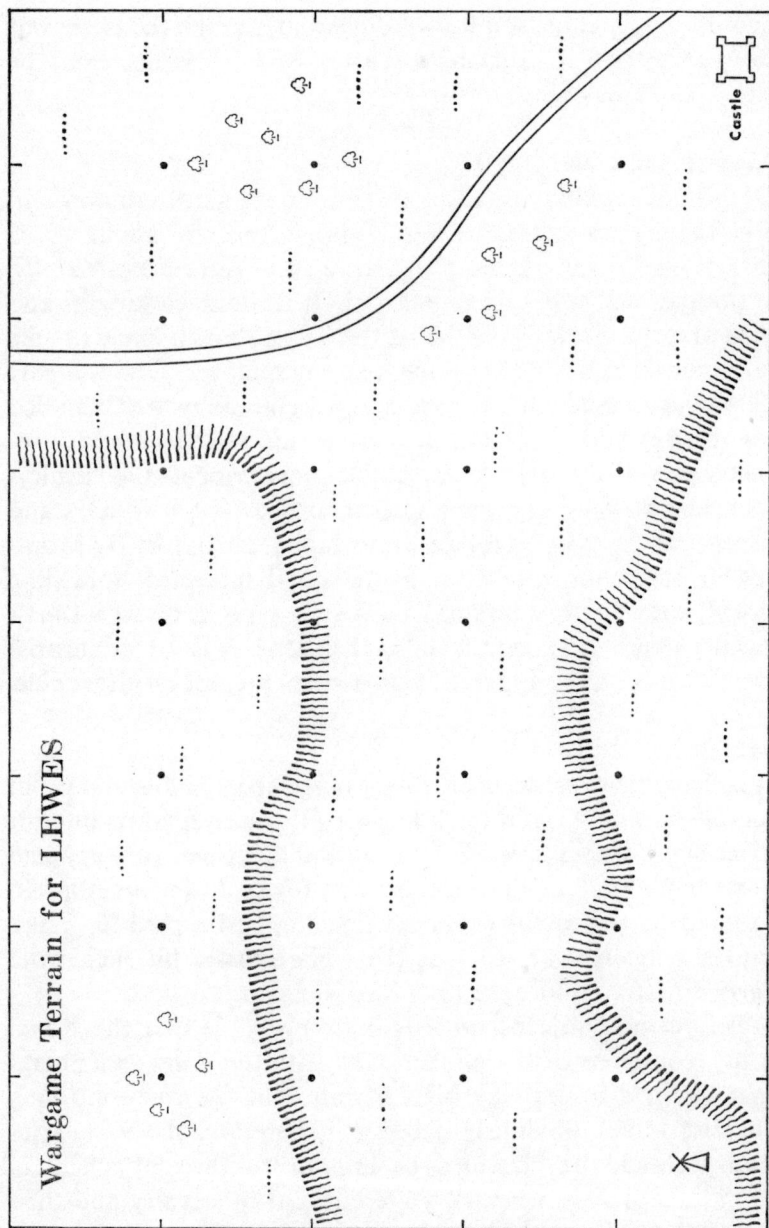

Wargame Terrain for LEWES

Castle

highest morale against the ineffective London levies, who were undoubtedly third class.

Terrain
In the early stages of the battle, during de Montfort's dawn advance, the undulating terrain afforded him the concealment that laid the foundations of victory. Otherwise the terrain had no influence on the battle, and was seemingly normal slightly undulating English countryside, with some undergrowth and clumps of trees.

Military Possibilities
1   De Montfort does not achieve initial surprise, and encounters a much larger army already formed up in a defensive position. Similarly, lack of surprise probably would have meant that Prince Edward and his cavalry were not first in the field and, although the cavalry would almost certainly have charged the Londoners on their front, they might not have pursued them so fruitlessly.
2   Edward and his cavalry do not pursue, and de Montfort's tactics have to be modified.
3   The King's infantry holds off de Montfort's infantry and cavalry until the Prince returns with his cavalry force. If such a Military Possibility is to be seriously considered when simulating this battle, consideration must be given to the difficulty of rallying scattered cavalry after a long pursuit. If it was difficult in the Peninsular War over 500 years later, it must have been almost impossible with the less-disciplined and highly impetuous cavalry of the feudal era.

9

The Battle of Bannockburn
1314

IN JUNE 1314 King Edward II, the unwarlike son of a warlike father, invaded Scotland with the objective of relieving Stirling Castle, whose English defenders had agreed to surrender to the Scots if not aided by Midsummer Day. Robert Bruce, King of Scotland, attempting to intercept the English army, elected to hold a partly open plateau overlooking the marshy valley through which the Bannockburn meandered to join the Forth. His right flank was protected by the burn at the Newmiln bog and the left by a forest on Gillies Hill. The Scots right wing was further protected by 3ft deep pits with a stake in each, lightly covered with sods and branches. Holes (or 'pottes') were dug in the road that ran behind them and iron calthrops with sharp points were strewn over areas where cavalry might be expected to charge.

Bruce divided his army of 8,000 to 10,000 men into 4 'battles' – the right commanded by his brother Edward, the left by Douglas and the young Stewart of Scotland, the centre by Thomas Randolph, the veteran Earl of Moray, and the reserve under his own command. With the exception of 500 cavalry under Sir Robert Keith, the Marshal, which were stationed about ¼ mile forward of the left wing, the knights were ordered to fight on foot like heavily armoured infantrymen.

On the afternoon of 23 June Edward's army came into view. It consisted of at least 3,000 heavy horse, 5,000 archers and 15,000 infantry. The cavalry was divided into 10 'battles' in 3 lines of 3 battles each, with the tenth forming an advance

The Battle of BANNOCKBURN 1314

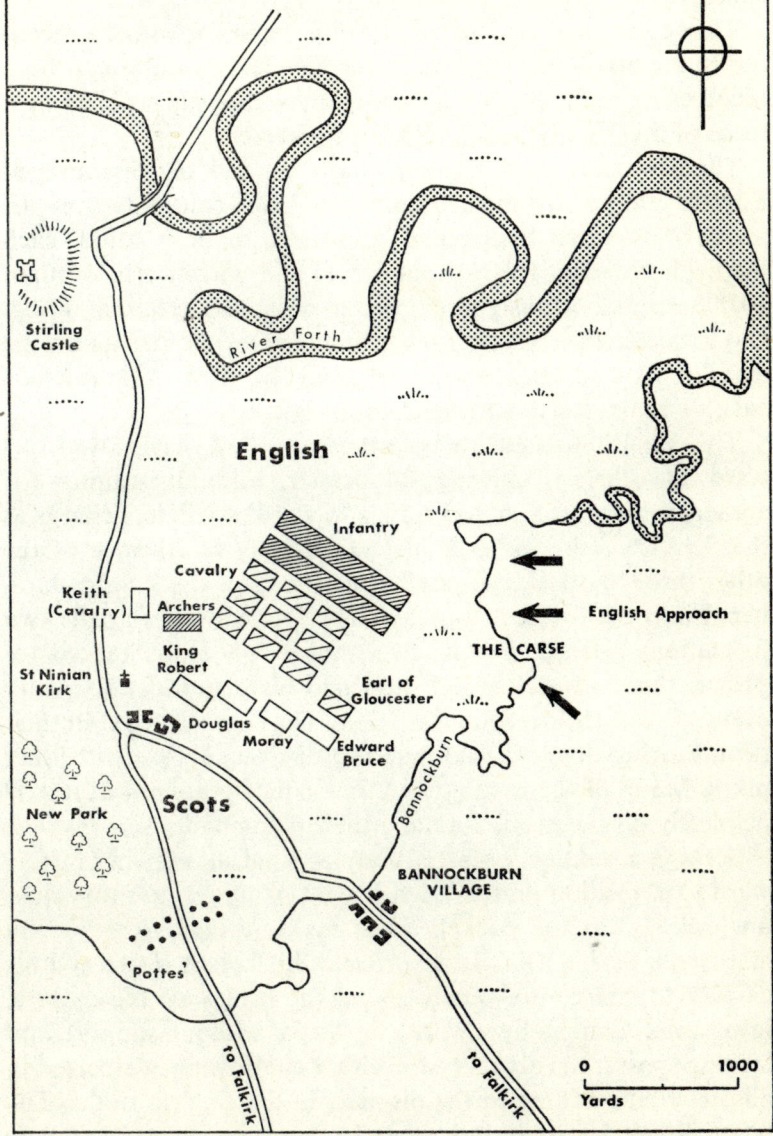

Stirling
Castle

River Forth

English

Infantry

Cavalry

Keith
(Cavalry)

Archers

King
Robert

St Ninian
Kirk

Douglas

Moray

Earl of
Gloucester

Edward
Bruce

THE CARSE

English Approach

Bannockburn

New Park

Scots

BANNOCKBURN
VILLAGE

'Pottes'

to Falkirk

to Falkirk

0   1000
Yards

F

guard. It was an impressive array, under gaily coloured pennons and banners, burnished arms and armour glittering in the sunshine.

Two companies of English cavalry came forward to reconnoitre the Scots position and, after Sir Henry Bohun had been killed charging at Bruce, one company was easily repelled by a force of 500 tightly packed Scottish pikemen.

There was no more fighting that day, and the discouraged Edward ordered his men to build a wooden causeway over the marshland so that the packed infantry formations could reach the field of battle. Throughout a cold and uncomfortable night 20,000 men laboriously attempted to cross the stream and bog. Daybreak saw the English main body across the stream on the marshy flats milling about in disorder. Only Gloucester's cavalry vanguard was formed up for battle.

This situation so encouraged the Scots that they moved forward in echelon, causing Gloucester, without waiting for archers to support him, to lead his 'vaward' uphill in a charge at the Scottish right wing, which was slightly in advance of the other three 'battles' The ponderous steeds of the knights lumbered into the forest of outstretched Scottish pikes as the two formations clashed. For a few minutes they stood locked together, then Gloucester fell dead and his demoralised cavalry drew off. But their respite was short. The Scots charged in their turn, bearing down the floundering English soldiers with their pikes. Many of the cavalry were immediately unhorsed, to roll helplessly on the ground among their plunging horses.

Seeking a vantage point, a body of English archers ran forward to a position on the English right flank, whence they fired unchecked into the packed Scots ranks, bringing down man after man. Seeing this, Bruce ordered Sir Robert Keith and his cavalry to charge diagonally round the fringes of the morass, into them. Caught by surprise in flank, without support and lacking spears, the archers were either cut down or dispersed in all directions, adding to the disorder in the English ranks. The Scottish cavalry so intimidated the English archers that they spent the rest of the battle firing flights of arrows from the rear over the heads of their own troops, doing little damage to their unseen targets.

The 9 remaining English cavalry divisions now came lumbering up the slope towards the Scottish centre and, as the huge mass of horse and foot locked together, the battle developed into a confused mêlée between Scots pikemen and English men-at-arms. In such tight order and on such a narrow front only the foremost English ranks could strike the enemy, those in the rear being unable to move. Mounted knights and men-at-arms, in small bodies, made ineffective charges that all failed to break through the pike formations. So it came about that the English archers and spearmen waited in disarray behind their struggling cavalry until the piles of their own dead were so high that any forward movement was impossible. The English formations began to falter, and the tidal wave of defeat mounted when thousands of Scottish camp followers, who had been watching the fight from Gillies Hill, swarmed excitedly down the slopes waving banners and shouting 'Slay! Slay!' Imagining the rabble to be Scottish reinforcements, the English wavered and fled. Soldiers in the rear who had not even struck a blow stared incredulously at comrades fleeing past them, until panic took hold of them too and they ran for their lives, turning the defeat into a rout.

Down the hill came the Scots, driving the English into the bogs of the burn, where they were smothered or drowned, and soon the narrow ravine was choked and bridged over by the slain. Together with some of his knights, King Edward turned tail and galloped off to Stirling Castle, but was refused admittance and struggled on to the Castle of Dunbar. Behind him Stirling surrendered.

Reconstructing the Battle
Begin the reconstruction at daybreak, with the Scots army divided into 4 battles on its slightly raised plateau, and the Scots cavalry forward of their left wing. The English force, with Gloucester's cavalry vanguard (about 300 men) in formation and under control, have about two-thirds of their army in a state of

disorganisation on the Scottish side of the marsh. The battle proceeds in its historical sequence with the Scots advancing (this is made easier by pre-mounting the figures upon an adhesive tile cut to the shape of the formation) and Gloucester's vanguard charging their right. A normal wargames mêlée takes place, with the Scots pike formation being advantageously placed because of its cohesion and power. The English archers move out to the flank and begin to inflict casualties upon the massed Scots ranks before being scattered by the Scots cavalry reserve, whose unseen angled approach is difficult to simulate on the wargames table because no self-respecting wargamer is going to permit his archers to be firing in one direction while a large body of cavalry is coming up on their flank! This can be solved by Chance Cards coordinating the cavalry's unseen approach with the archers' preoccupation with the main battle. Once this problem has been settled, the archers will have little chance when attacked by cavalry, so history should be repeated.

Local rules will handicap the tightly packed English, eliminating the shock effect of cavalry, which, unable to back off from the menacing spear points, cannot come in again with a rush. It is unlikely that figures will be required to represent the Scottish camp followers whose spontaneous downhill surge caused the final English rout, nor will it be necessary to provide model soldiers for the entire English army, as only part of the cavalry and about 5,000 archers ever got into the battle. In reality 10,000 Scots defeated 6,500 English.

Commanders' Classification
Bruce and all his subordinates must be classified as Above Average. King Edward and his commanders, poor tacticians as they were, are undoubtedly Below Average.

Quality of Troops and Style of Fighting
As an army, the English should have been superior but, on the day, inept leadership, the terrain, fate even, made them inferior to the Scots. The English commanders assumed that Bruce would not attack because he had taken up a defensive position

on the hill, and that his tightly packed formations would be decimated by the English archers until they were so weakened that they could be ridden down by heavy cavalry. But this was not the case and the English were defeated because by 1314 the heavy horseman was no longer master of the battlefield; even brave and determined cavalry could not defeat steady pikemen unless supported by archers, who themselves required the backing of heavy troops. Gloucester's cavalry vanguard was too tightly packed to be effective, and its impetuous charge showed a complete and disastrous lack of respect for the Scottish pike formations. Bruce knew that his formations, on suitable ground, could hold their own and resist cavalry charges, although he was aware that the few cavalry in his numerically smaller army could not cope with the English horsemen.

The ill-conceived night march prevented the English from being in any sort of battle order at daybreak, so that they were an easy prey to Bruce's advanced echelons, and were never able to employ even half their numbers. In this truly feudal conflict the English archers, still 30 odd years away from inspiring awe and terror in their foes, were left to fend for themselves, and after being ridden down quite naturally showed reluctance to move without support again. The Scots pikemen were superb. It is instructive to realise that their schiltrons or 'battles' were the foundation on which later commanders built their infantry 'squares' for holding off cavalry, in use even as late as Waterloo in 1815.

Morale
Their successes of the previous day and the sight of the disorganised English at daybreak raised the Scots' morale. The English began the battle with low morale, made worse by the course of the conflict.

Terrain
Bruce made the terrain an ally of overwhelming strength, justifying 'King Robert's Testament': 'Always fight on foot, positioned on a hill and flanked by woods, with a marsh before

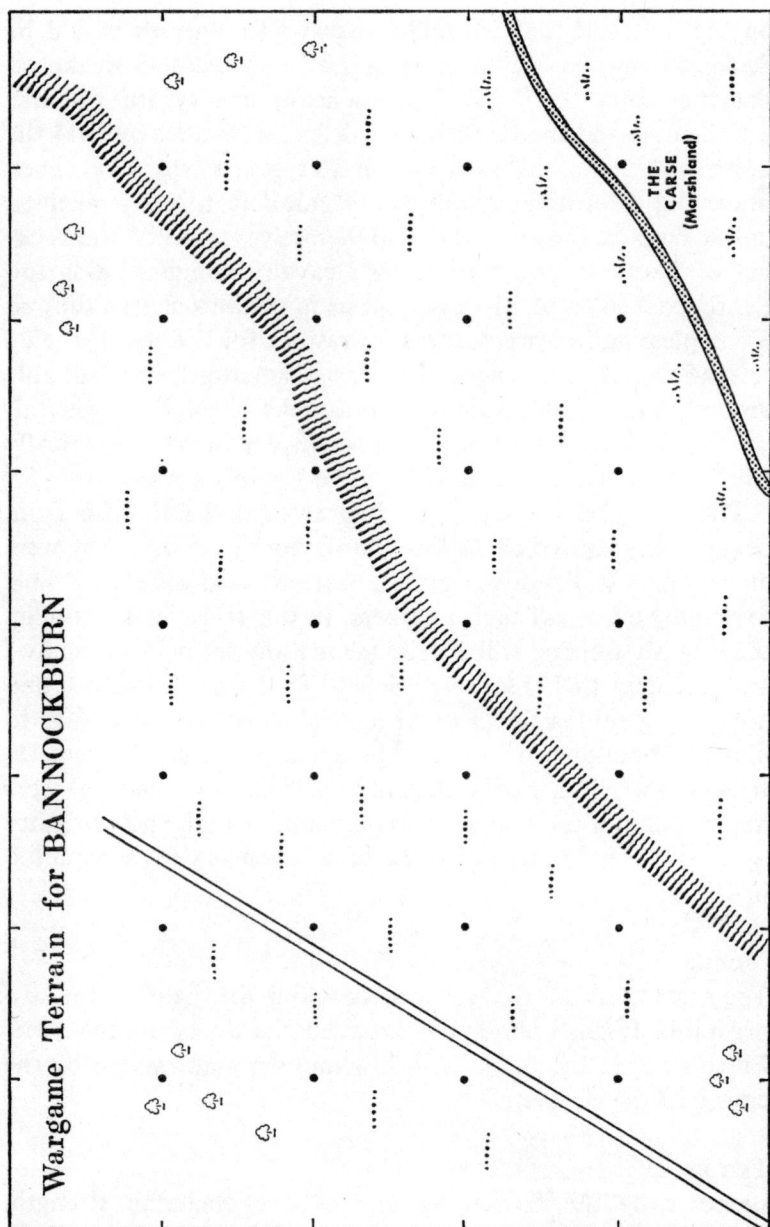

Wargame Terrain for BANNOCKBURN

THE
CARSE
(Marshland)

you; keep the battle-front narrow; lay waste to the land through which the enemy must pass; keep your foe awake all night by noise and mock attacks.'

Military Possibilities

1   The English achieve greater success on the previous day, so that their subsequent tactics are pursued with greater confidence, and those of the Scots are tempered by caution.

2   The English night march is performed more efficiently, so that their formations are organised at daybreak.

3   Gloucester's charge is more successful, dispersing or halting the Scottish right.

4   The English archers turn back the Scots cavalry, as they had done in past and future battles against the Scots.

10
The Battle of Morlaix
1342

MORLAIX, THE first pitched land battle of the Hundred Years War, was won by the English because they used their success against the Scots at Halidon Hill in 1333 as a rehearsal for this and almost all the other great battles to follow. In September 1342 an English army of 3,000 under the Earl of Northampton, besieging Morlaix, was suddenly threatened by Count Charles de Blois and a relieving army of 15,000 to 20,000 foot and horse. Abandoning the siege, Northampton marched along the road to Lanmeur to find a position to make a stand. He found it 4 miles from Morlaix – a small ridge across the road with a forward slope some 300yd long. A wood to his rear, ideal for hiding his baggage wagons, also prevented him from being out-flanked by cavalry. The English formed up in a line 600yd long and 50yd from the wood, with dismounted men-at-arms in the centre and archers on the flanks, and dug a trench as an obstacle to cavalry in front of their line.

In 3 successive columns, the leading one formed of dis-mounted local levies, the French advanced straight at the wait-ing English. A hail of arrows reduced the first column to disorder, and the second, of mounted men, was stopped by the hidden trench, where a confused mass of men and horses piled up, an easy target for the English archers. A few horsemen suc-ceeded in remounting and crossing the trench, but they were all brought down by arrows.

Now followed a quiet pause while the surviving French re-tired and the third column prepared to advance. It was larger

The Battle of MORLAIX 1342

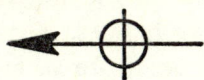
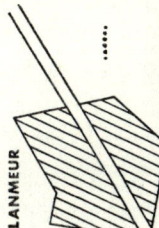
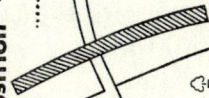

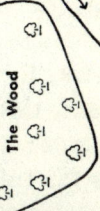
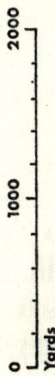

LANMEUR

French Approach

English Position

The Wood

Brook

to Morlaix

4 miles

0   1000   2000
Yards

than the whole English army, extending beyond the English
wings and threatening their position from both sides. The 2
leading columns, each greater than his own small force, had
been repulsed, but Northampton knew his archers to be short
of arrows, though they had recovered all they could; and the
trench was now filled with bodies and presented little obstacle
to determined men.

As the third column began a ponderous advance on foot,
Northampton withdrew his men to the shelter of the woods,
forming a defensive line along the edge of the trees. The archers
reserved their scanty ammunition until the French came close,
and then fired sparingly but effectively. Not a single Frenchman
penetrated the position at any point, although they swung round
the flanks and almost surrounded the wood.

In fact Northampton's worries were minimal compared to
those of Charles de Blois, whose men, including a number of
Genoese mercenary crossbowmen, were deserting on all sides.
With night approaching, he abandoned the contest and withdrew slowly to Lanmeur, at which Northampton formed his
small band into a defensive formation and left the wood to
return to the siege of Morlaix.

Reconstructing the Battle
Like Taginae, this was strictly an affair of firepower, with not
a single instance of a blow being struck in hand-to-hand combat.
As long as the rules give due weight to the effectiveness of the
English archers, the events of 1342 cannot but be repeated on
the table-top battlefield. The lowering of morale caused by
losses, together with the already poor morale of the French
local levies, will ensure this. The French contributed to their
own downfall by attacking in 3 separate divisions, so that never
more than 5,000 or 6,000 of them at any time threatened an
English force, 3,000 strong, with extremely high firepower and
operating for some of the action from concealment.

As the French fought in successive 'battles' it is not necessary
to provide figures for all of them (though the purist wargamer

may cavil at using local levies to represent men-at-arms, and vice versa).

Commanders' Classification
Northampton, on the day, was well Above Average, one of the best commanders in this book. Charles de Blois was well Below Average, however, and one of the worst.

Number and Quality of Men, and Style of Fighting
The total force besieging Morlaix is said to have been 3,000, comprising 2,000 men-at-arms and 1,000 archers, which means that the French divisions were turned back by only 1,000 men. De Blois led an army of 15,000 to 20,000 foot and horse, split into 3 large columns, each of 5,000 to 6,000 men. The leading column was formed of dismounted local levies, ill trained and of low morale; the second was formed of mounted men; and the third of armoured and dismounted men-at-arms.

The French had with them a body of Genoese mercenary crossbowmen, who were expensive professionals, but it is unlikely that they amounted to more than 500. They did not play a very prominent part in the battle, perhaps because, as at Crécy, they were not given very much chance to do so, or because, as they advanced, they were overwhelmed by the weight of archery fire turned upon them.

The result of the Battle of Morlaix failed to convince the French that it was almost hopeless for cavalry to make a frontal attack on a position defended by men-at-arms supported by archers. Alone the superbly trained English archers turned back the 3 successive French columns, but had one or other of these columns come to grips with the English (as at Agincourt), it is quite likely that they would have been driven back by the combined efforts of the dismounted men-at-arms (well trained professional soldiers) and the nimble archers, now laying about them with sword and maule. The French local levies were frightened and unwilling peasants, probably only armed with sticks, clubs or scythes. The mounted French in the second column would have been heavily armoured knights or men-at-arms, professional soldiers with training, discipline and skill, though their headlong charge at the enemy was now outdated. The third column was composed of similar professional soldiers, though this time they fought on foot.

The French military system in the mid-fourteenth century was the epitome of feudalism, with its lords and their retainers jealous of each other and unwilling to cooperate. Tactical control, therefore, was almost impossible. The mounted knight and man-at-arms considered themselves the masters of the field, and the infantryman was scorned and inadequately employed. Forced to attack by the English method of waiting in carefully chosen defensive positions, the French knights were immobilised when their horses were killed by the archers; and when they attacked on foot, their infantry, equally vulnerable to arrows, reached the English position too exhausted to be effective.

Morale
Made aware of their potential at Halidon Hill some 9 years earlier, the English archer/men-at-arms combination, led by exceptional commanders, produced the highest possible state of morale. The French, superior in numbers and seemingly the pursuers, were probably of equally high morale at the onset of the battle, except for their peasant levies whose morale must inevitably have been low.

Terrain
As in almost every battle of the Hundred Years War, the English position in Morlaix was all important. It was a copybook defensive position, suited to the weapons of the day, and made even more impregnable by Northampton's retreat into the wood, whence his men could fire from concealment.

Military Possibilities
1   The French catch the English before they have taken up a defensive position. This is unlikely, because the latter were able to dig a trench, indicating that Northampton had reached his selected position with time to spare.

Wargame Terrain for MORLAIX

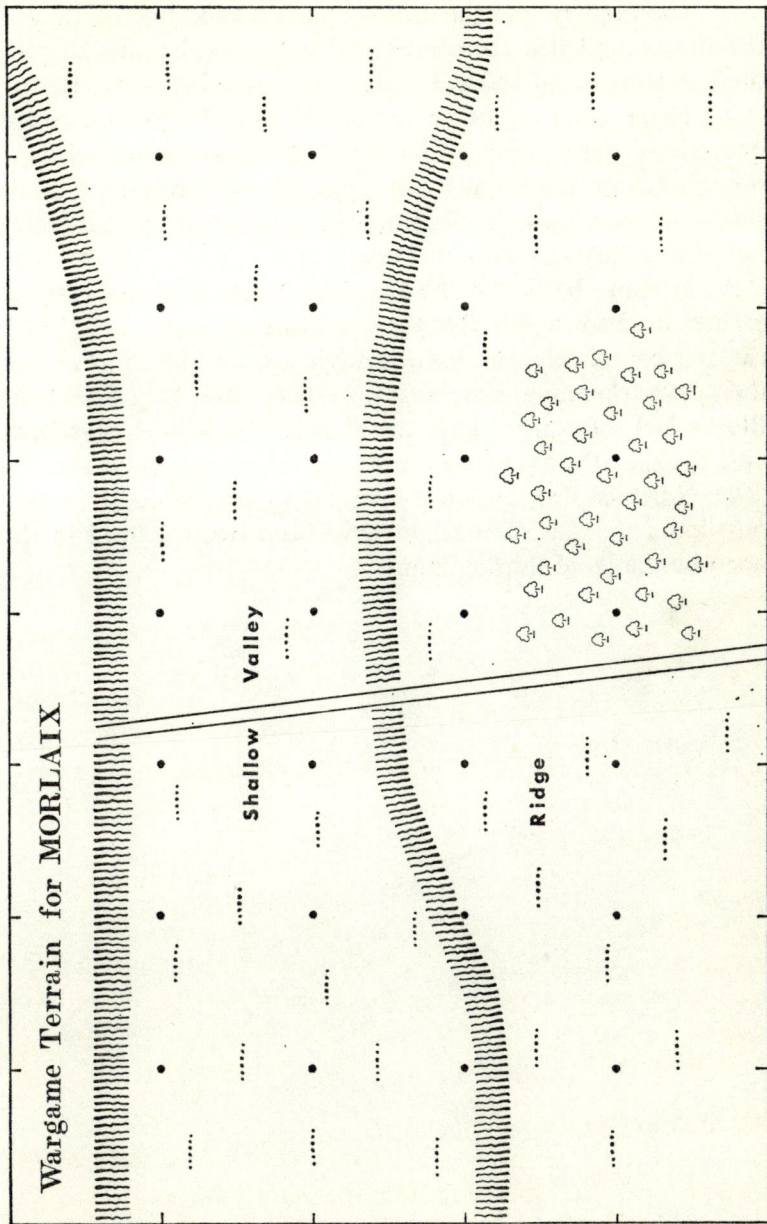

Shallow Valley

Ridge

2   The French alter the order of their attack, first sending in the dismounted men-at-arms backed by the cavalry, and keeping the unreliable local levies in reserve for mopping-up purposes.

3   The French attack en masse, sending the cavalry round one side of the English position and the local levies round the other, so as to stretch the small English force. Even when in the shelter of the woods, the English were still vulnerable to infantry infiltration through flank and rear.

As in many battles of the Hundred Years War, the English archers fired so rapidly that they ran short of arrows. Had they run out completely, the French might have come to grips and their overwhelming numbers prevailed. But the Count de Blois's lack of control indicates that he would not have been able to take advantage of even such a favourable situation.

4   Northampton does not retreat into the wood. There is a possibility that his men might have been overwhelmed in the open by the large third column.

11
The Battle of Auberoche
1345

IN SEPTEMBER 1345 the Earl of Derby made a successful raid into Gascony to capture the Castle of Auberoche, which he garrisoned before returning with prisoners and booty to Bordeaux. Built on a rocky prominence overlooking the River Auvezere, the castle was of some strategic importance as it dominated and blocked the narrow valley, which was lined with meadows and bordered on each side by heavily wooded slopes. Count de l'Isle reorganised the defeated French army and, accompanied by siege engines, returned to besiege Auberoche, setting up his main camp in the valley, with a lesser camp in a still narrower valley north of the castle.

Warned of the siege, Derby quickly gathered together a scratch force of 400 men-at-arms and 800 archers, and marched them 50 miles, mostly under cover of woods, until, by the evening of 20 October 1345, he and his men were bivouacked in a wood some 700yd from the main French camp. The greatest precautions were taken to ensure that their presence was not revealed, as they awaited the arrival of Pembroke with re-inforcements. But as these had not arrived by the afternoon of the next day, Derby, though only able to muster 1,200 men against 7,000 French, decided to attack. A personal reconnaissance revealed that the unsuspecting French were cooking their evening meal.

The English leader decided to attack across a level approach, practicable for horsemen, some 300yd south of the main French camp, which could be reached by a woodland track.

The Battle of AUBEROCHE 1345

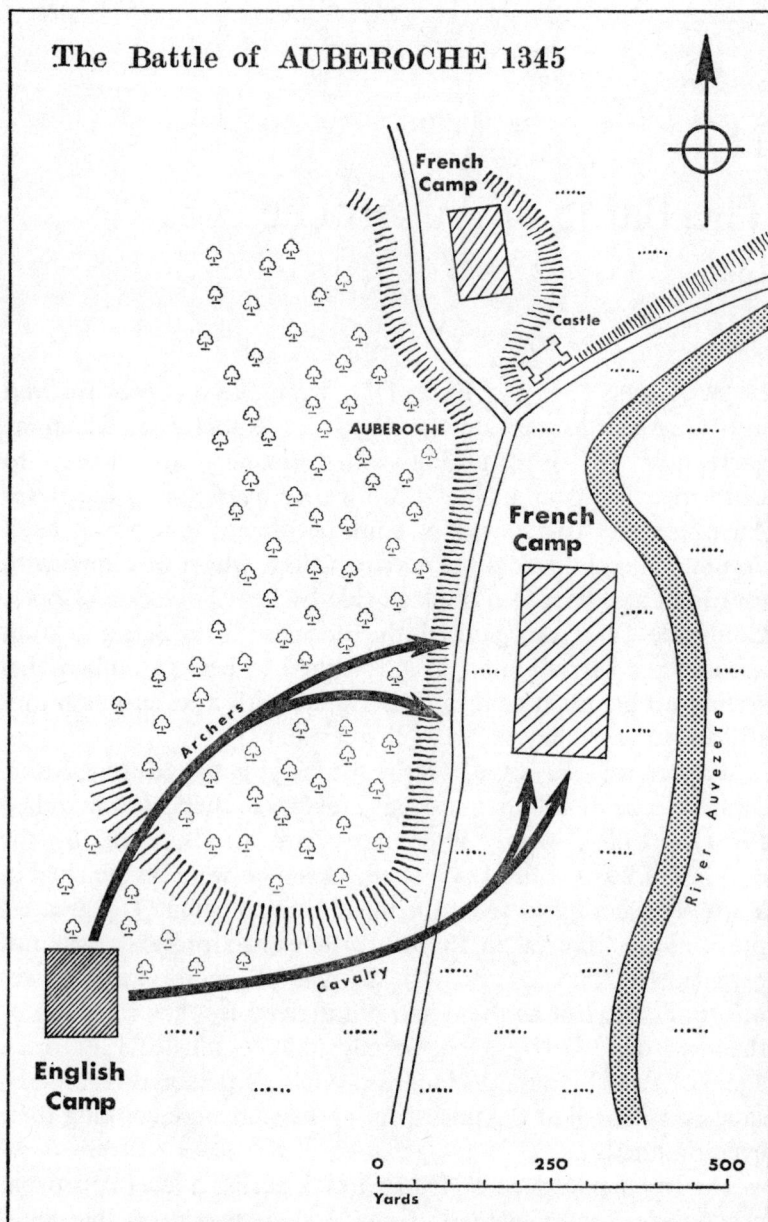

French
Camp

Castle

AUBEROCHE

French
Camp

River Auvezere

Archers

Cavalry

English
Camp

0   250   500

Yards

The archers were to creep through the undergrowth to line the edge of the wood at a point exactly opposite the French camp, whence they could give supporting fire to the charging cavalry. When the cavalry reached the camp and masked their fire, the archers were to switch their aim to any other available targets. The operation required exact timing, and Derby's orders were necessarily precise.

At his signal the attackers moved silently forward to take up their allotted positions. Smoke rose from the French campfires, by which unarmoured men took their ease. Suddenly and shockingly their peace was shattered by shouts of 'Derby!' and 'Guyenne!' as the archers fired from the woods and the cavalry galloped from their place of concealment to charge across the 300yd of meadow that lay between them and the French tents. Confusion reigned in the French camp as the soldiers tried to arm themselves.

The first flights of arrows caused great casualties to the crowded French, and the horsemen, following up, found little opposition. When they reached the centre of the camp, the horsemen disappeared from the sight of the archers, who switched to another target – an attempted French rally under some officers who had managed to struggle into their armour and unfurl their banners on the outskirts of the camp. The small groups of men who tried to join them were quickly scattered.

Although Derby had been unable to convey any signal, Sir Frank Halle, commanding the beleaguered garrison in the castle, had seen the attack, and now joined in, with every available man mounted, to charge the French camp from the north. The sally was supported by archers firing from the ramparts, though at long range. The sudden eruption of Halle's little cavalry force broke the remaining French resistance, the few survivors fleeing the field. Throughout the battle the French in the small camp north of the castle had made no move, and they fled at the same time as their comrades from the main camp.

This complete and astonishing victory saw Count de l'Isle and the flower of the chivalry of Southern France killed, wounded

G

or captured. Although one of the lesser battles of the war, Auberoche established the same moral supremacy in Gascony that was set in Brittany at Morlaix in 1342 and in Picardy at Crécy in 1346. The French later learned their lesson under Bertrand du Grueselin, and forgot it again at Agincourt; and not until the last decade of the Hundred Years War did they engage British forces in the field with any hope of success.

Reconstructing the Battle

This, the smallest and perhaps the most stirring battle in the book, is best simulated by the French 'Commander' believing that he is besieging a castle, and then confronted with attack by archers and a cavalry charge. The scene should be set as it was on the evening of 20 October 1345, with the French in two camps and the wood to the south, from which Derby's cavalry attacked, on the fringe of the wargames table, so that the cavalry are not in evidence; nor are the archers placed in position on the table, as they are concealed in the woods and undergrowth along the west side of the French camp.

In this semi-feudal age it was exceptional for a commander to have sufficient control over his force to carry out a simultaneous two-pronged attack, and it might make the French 'Commander' happier if Derby's coordination is controlled by Chance Cards. The odds should be so slanted as to allow Derby to reproduce his historical feat, but allowances made for the cavalry to attack before the archers are ready, or the archers to open fire before the cavalry are ready to advance. As soon as the first volley of arrows has been fired and the charging cavalry come into sight, it will be necessary for the French in the south camp to test their morale. Even those who stand must fight at a disadvantage because of (a) their surprise and fear, (b) their disarray and lack of defensive formation, and (c) because a large percentage of them will be without armour or weapons. An agreed percentage, therefore, will fight the battle without armour, so lowering their fighting ability (this can be simulated in the same manner as at Stamford Bridge, p 68). Another agreed proportion will have to move measured distances from wherever they might be at the onset to their tents, in order to arm themselves.

Commanders' Classification

Derby revealed himself as an Above Average commander on this occasion and de l'Isle, his ability gravely affected by force of circumstances, must be classified as Below Average. Sir Frank Halle, commander of the beleaguered garrison, also displayed sufficient initiative and courage to be graded Above Average.

Number and Quality of Men

Derby led a small but well-balanced force of 400 mounted men-at-arms and 800 archers. De l'Isle commanded a force of 7,000 French which, for want of more precise information, is deemed to consist of 2,000 mounted knights and men-at-arms and 5,000 dismounted men-at-arms – split into 5,000 in the main camp and 2,000 in the smaller camp to the north. The size of Halle's garrison is not known but it is reasonable to assume that his force consisted of about 100 archers and 100 men-at-arms, whose horses were stabled within the castle.

The English archers were the outstanding soldiers of their day and the men-at-arms were well armed, trained and disciplined. Cool and competently led, coordinating their respective roles, both archers and men-at-arms displayed a most workmanlike aggression. The French men-at-arms were also trained and disciplined soldiers, but they could only fight a desperate defensive battle.

Morale

Stimulated by their warrior king, Edward III, the English soldier of this age was sublimely confident in both himself and his leadership, so that his morale was as high as it could possibly be. It is possible that this arrogance depressed French morale, though there is no reason to suppose that the French, besieging Auberoche in overwhelming numbers, should be less than first class *before* being attacked by Derby's force. Their morale must

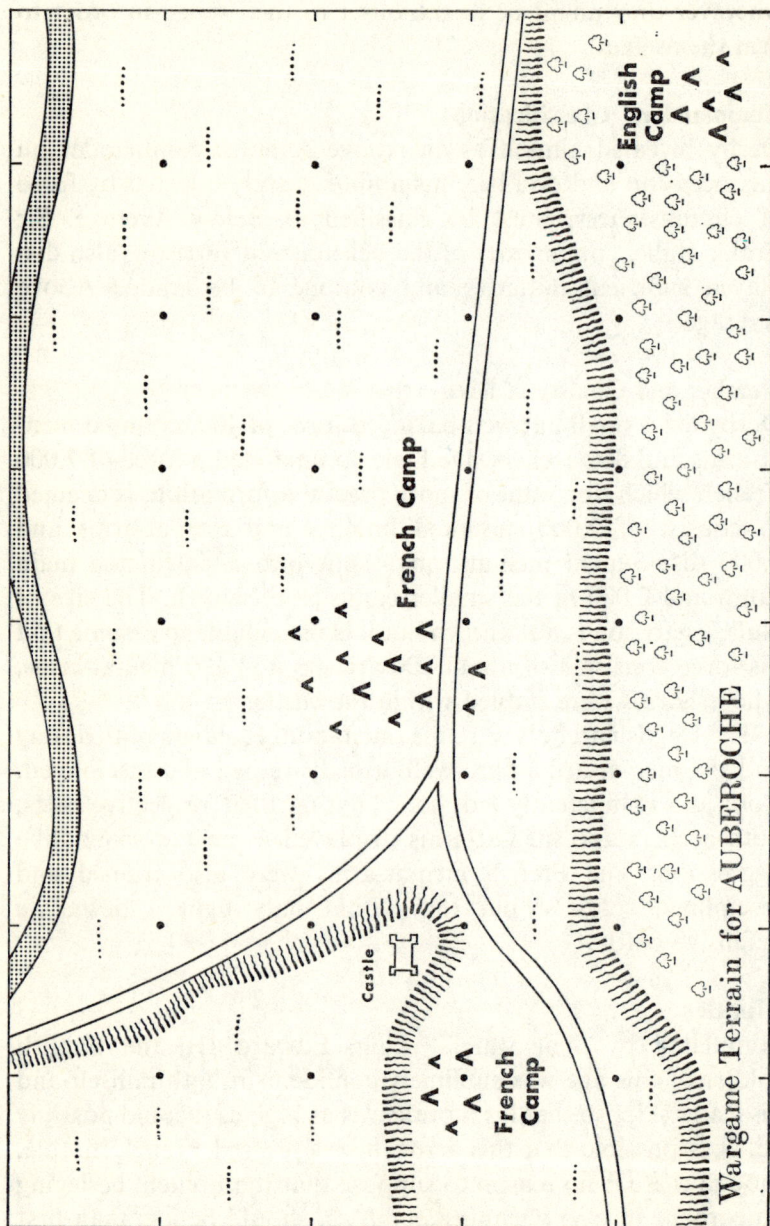

Wargame Terrain for AUBEROCHE

English Camp

French Camp

Castle

French Camp

have plummeted on the English onslaught, however, and the course of the action did nothing to improve it.

Terrain
The terrain decided Derby's victorious tactics. Without the cover afforded by the woods and undergrowth, his small force would never have been able to approach the enemy unseen.

Military Possibilities
1 The French discover Derby's approach (a) in its early stages or (b) when his force is forming up on the edges of the wood for attack. Either of these eventualities would completely change the trend of the battle, leaving the French with over-whelming numerical advantages.

2 The French rally, to organise a stronger resistance.

3 The French troops in the north camp move down to aid their comrades or prevent Halle's force from sallying from the castle.

4 After their opening volleys, the English archers' fire is sufficiently masked by their own men to reduce its effectiveness.

5 Pembroke arrives with reinforcements in time to take part in the attack, a possibility which could be combined with No 1.

12

The Battle of Neville's Cross
1346

FOLLOWING HIS defeat at Crécy and the melting away of his army, Philip of France sought to relieve the pressure on him by persuading David II, King of Scotland, to invade England. David succumbed to the temptation, and in October 1346 marched an army of some 18,000 veteran Scots troops and French auxiliaries, including 1,000 cavalry, over the border into England, being assured that Edward and his chief commanders were absent so that 'here are none to oppose our progress save churchmen and base artisans'. He crossed the Tyne at Ryton, above the town of Newcastle, and advanced into County Durham to encamp, on 16 October, at Beaurepair (Bear Park), about 2 miles north-west of the city of Durham.

Within the city itself consternation prevailed, for it seemed to be at the mercy of the invaders; but its position was not as bad as it appeared, and an English army was swiftly assembled. It numbered about 16,000 well-equipped men-at-arms, archers and infantry, led by the northern barons – Ralph, Baron Neville of Raby; Henry, Baron Percy of Alnwick; Musgrove; Scrope; Hastings; and the ubiquitous Edward Baliol.

The English force advanced cautiously eastwards, moving slowly by the Red Hills on the west of the city of Durham until it reached the ground on which the forthcoming battle was to be fought. Lying west and west by north of the cathedral, the battleground was a level area but dipping on the west and falling away steeply towards the river on the east. The Scots right flank

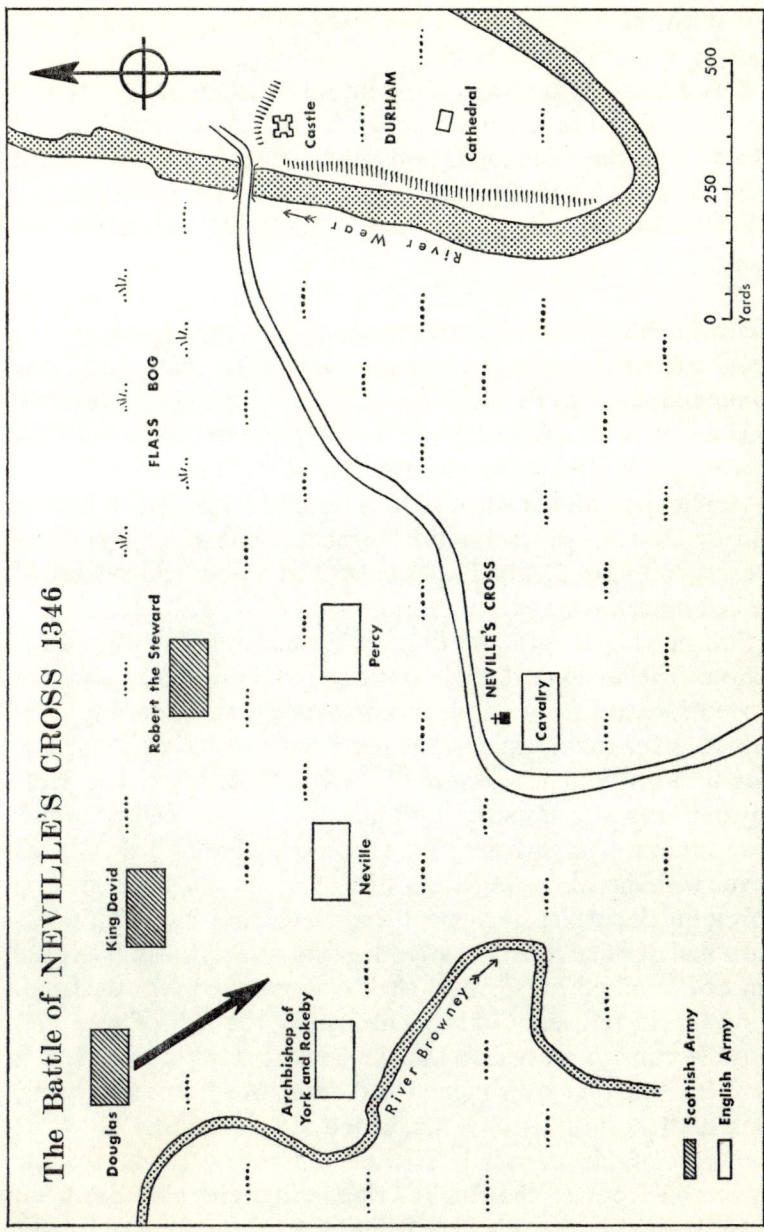

The Battle of NEVILLE'S CROSS 1346

Douglas

King David

Robert the Steward

Archbishop of York and Rokeby

Neville

Percy

FLASS BOG

NEVILLE'S CROSS

Cavalry

River Browney

River Wear

Castle

DURHAM

Cathedral

Yards
0   250   500

Scottish Army
English Army

rested on the deep ravine of the River Browney, which curved sharply across Douglas's front.

David formed the Scots army into 3 divisions. The first was led by the High Steward of Scotland, the second by the Earl of Moray and Sir William Douglas of Liddesdale (then named 'The Flower of Chivalry'), and the third division, consisting of select troops and a party of French auxiliaries, was led by the King in person.

The English were disposed so that Lord Percy led the vanguard, which in the battle became the right wing, opposed to the Scots left wing under the High Steward. The main body was commanded by Lord Neville and, as the centre, in the battle joined issue with the Scottish main body and centre under King David. The English rearguard (the left wing) under Rokeby was in conflict with the Scottish right wing led by the Earl of Moray and by Douglas. Each English formation threw out a body of archers to its front. The English also had a powerful reserve of mailed horsemen, under the command of Edward Baliol.

Still moving slowly, the English advanced and deployed for action. The Scots left their position on Durham Moor and moved forward to meet them, immediately encountering difficulties as the ravine narrowed their front and forced Douglas's men into the flank and centre of the King's division. The Scots advance was also impeded by walls and hedges, behind which were stationed English archers, whose arrows played their usual havoc with the advancing Scots. Sir John Graham, remembering how a quick cavalry movement against the archers at Bannockburn had decided the day, asked leave to attack them. 'Give me but one hundred horse and I shall disperse them,' he declared. King David refused. Graham, furious at the loss of men and sensible enough to realise that archers at 'long bowls' had a terrible advantage over men armed with sword, axe and spear, took matters into his own hands and struck the first blow. At the head of his personal followers he rode straight for the archers in front of the English right wing, charging down on them so quickly that his little band broke through in one place and dispersed the archers there. At short range, however,

Graham's horse was shot down and he was wounded, but he managed to regain the Scots lines.

The High Steward, quickly grasping the situation, ordered his men to charge the partly disordered English right wing. Momentarily freed from the nagging arrows, the Scots came on with such impetuous fury that by sheer weight of sword and battleaxe they hurled the English column back in confusion against that of Lord Percy, whose wing was then in danger of rout. At this moment of crisis the value of possessing a cavalry reserve under a capable commander became apparent, for Baliol, with great spirit, charged the Scottish troops threatening Percy. Not only was the Scots attack on the right wing repulsed, but that repulse was converted into a complete rout, and soon the division of the High Steward was a bunch of fugitives. The High Steward did his best to re-form and reorganise his troops, who were entangled among hedges and ditches, and once again decimated by the fire of the now steady English archers.

The battle between the centres had been proceeding on almost equal terms, but Baliol, forgoing the temptation to pursue the High Steward's beaten division, wheeled his men and charged the open left flank of the Scottish King's division. Attacked in front by Neville, whose men had poured through gaps in the enclosures to charge in a somewhat confused but nevertheless desperate manner, and on the left flank by Baliol's cavalry, the Scottish centre slowly gave way. In spite of the King, surrounded by his nobles, fighting bravely, his division began to break up, the fugitives taking off towards the right, where Rokeby was doing more than hold his own. The flight of the centre, however, also destroyed the Scots right division, which, hampered by the nature of the ground, could not retreat; caught in enclosed fields and between hedges, the men of that division were slain without mercy.

On all sides the Scots had now given way, but their King, by his exhortation and example, repeatedly brought masses of them back to the fray. It was in vain and at last, as at Flodden almost 200 years later, the remaining knights formed a ring around their monarch and stood at bay. In spite of their gallant defence, at noon the Royal banner was beaten down, and, seeing it fall, the remnants of the Scottish army fled in despair. The 80 or so knights remaining around David surrendered, and at last the King himself was taken.

The English losses are not known, although in such a fiercely contested battle they must have been severe. The Scots undoubtedly lost more, perhaps half being slain on the battlefield and many more in the subsequent pursuit.

Reconstructing the Battle
Both armies are set out in their historical formations, with Douglas's men so positioned that the ravine will affect their progress. Graham's small cavalry charge could be controlled by Chance Cards that allow him to reach the English archers without being turned back by weight of fire. Similarly, Baliol's vital cavalry charge should be controlled by Chance Cards.

Commanders' Classification
Neville and Baliol seem to deserve the classification of Above Average. Other commanders on both sides are Average, no one earning a Below Average classification.

Number and Quality of Men, and Style of Fighting
The English army consisted of about 1,000 mounted knights and men-at-arms, 2,000 archers and some 12,000 infantry. The Scots army totalled about 18,000 men, including perhaps 1,000 mounted men-at-arms.

Composed of veteran troops and French auxiliaries, the Scots army may well have been of a better general quality than Neville's force, whose infantry undoubtedly included some ill-trained and poorly disciplined local militia. The cavalry on both sides were trained and experienced men-at-arms, supported by knights and men of quality. Both English, Scots and French infantry (except Neville's levies) were trained soldiers armed with swords and spears. Neville's great advantage was his force of archers, although those at Neville's Cross might not have been of quite the highest quality because, only 7 weeks earlier, 5,000 archers had fought at Crécy.

Wargame Terrain for NEVILLE'S CROSS

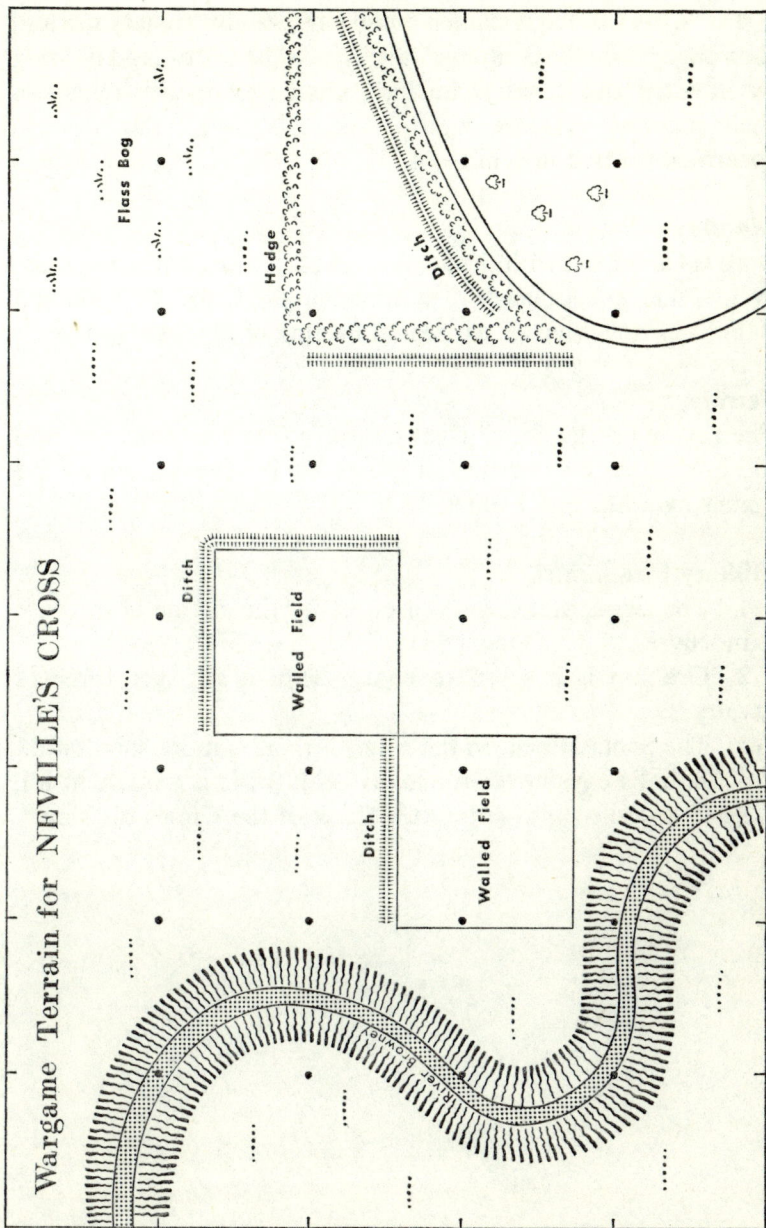

Flass Bog

Hedge

Ditch

Ditch

Walled Field

Ditch

Walled Field

River Browney

The Scots were first shaken and disordered by archery fire and then engaged in the contemporary style of hand-to-hand fighting by infantry and cavalry. Lacking missile men, they relied on their prowess in close fighting, probably with the Scottish spearmen formed in schiltrons.

Morale
Both sides started with the same level of morale, but subsequent events, such as Douglas's men impeding the King's division and Baliol's cavalry charge, caused a lowering of Scottish morale.

Terrain
The ravine on the Scots' right hampered their movements, and the English archers made full use of the cover afforded by the numerous walls and hedges.

Military Possibilities
   1   The Scots allow for the effect of the ravine upon their manoeuvres.
   2   Graham is allowed to charge with a stronger force of cavalry.
   3   The Scots attack on the English right is more successful.
   4   Baliol's cavalry charge is (a) held, (b) is not made at all, or (c) does not continue into the flank of the King's division.

13

The Battle of Shrewsbury
1403

SHREWSBURY WAS an important battle in English history, deciding the reigning family in England for the next 60 years. A singular feature about it was that 2 English armies engaged each other with the longbow.

In July 1403 Henry IV was on his way to Northumberland to join the Percys, who had done much to place him on the throne, in operations against the Scots; but Henry Hotspur, the Percy leader, was marching south through Lancashire and Cheshire, having formed a coalition with Douglas of Scotland and Owen Glendower of Wales to seize the throne. Hearing of this at Burton-on-Trent, Henry turned his army west and began levying men from the area.

Since Shrewsbury commanded the passage of the River Severn, it had been chosen as the meeting place for the rebel forces. Knowing that his son, Prince Hal, was in the town after raiding in North Wales, Henry marched quickly to reinforce him, reaching Shrewsbury on the night of Friday, 20 July, having marched 45 miles from Lichfield in a day. Hotspur arrived some hours later, after a march of over 250 miles, to find that Henry, with superior numbers, held the town, while his own Welsh allies were not to be seen. Hotspur was aware that it would be difficult for the 2 armies to unite without one or the other first obtaining possession of the town.

Among Hotspur's hastily assembled army of 10,000 men was Douglas with a force of Scots who had been captured by the Northern barons in the previous year at Homildon, where

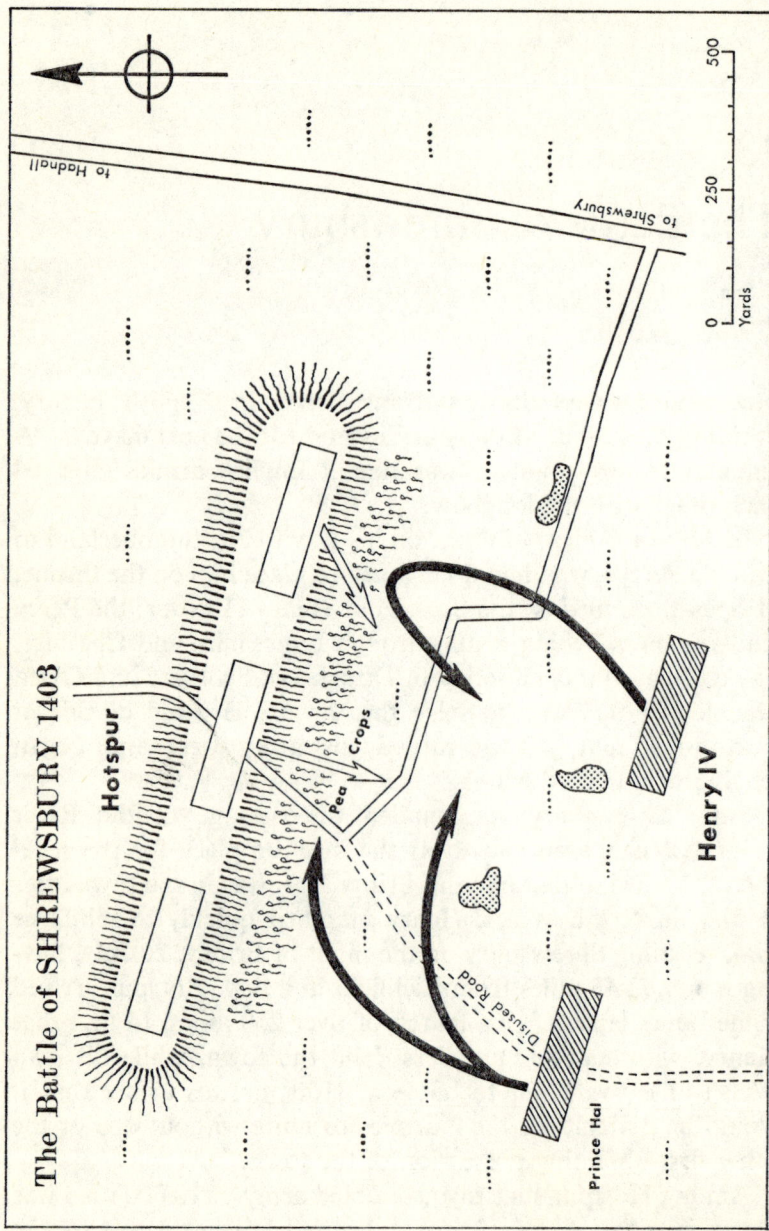

The Battle of SHREWSBURY 1403

Douglas had lost an eye. They agreed to serve Hotspur in lieu of ransom while grasping the opportunity of again fighting at least some English opponents.

Hotspur fell back some 3 miles northwards towards Whitchurch and formed his slender army into battle array on a ridge running east and west which stood out prominently in a countryside for the most part flat and featureless. Three ponds lay to his front, which was also covered by a field of fully grown peas, entwined to form a slight obstacle, though they were flattened long before the end of the battle.

About 12,000 strong, the King's army was drawn up in 2 divisions on a frontage of 800 to 900yd, commanded on the left by the Prince and on the right by the King.

The battle began with a fire-fight, Hotspur's Cheshire archers (there were none better in those days) causing great destruction in the ranks of the Royal archers, so that their fire became ragged and slow and they began withdrawing in some disorder. Encouraged, Hotspur's men-at-arms advanced down the slope and attacked the King's men, who had rallied and faced about, and a general mêlée took place around the ponds. On the left the Prince's division had not suffered such heavy casualties, although the Prince himself was wounded in the face. Nevertheless, he led his force in an angled charge that rolled back part of the rebel right wing; at the same moment the King's division (in column instead of the usual line) delivered an oblique attack up the slope between the ponds, forcing the rebels' left back so they became intermingled with their disordered right. Then Prince Hal's division pushed behind the enemy so that they were threatened from 3 sides – rear, right and front.

During the battle Hotspur and Douglas with 30 chosen knights, all swearing to kill the King, hewed their way into the very heart of Henry's position; but Henry had changed his armour and several other knights were dressed in the Royal colours. Douglas killed 4 of them, only to see the 'King' arise again on each occasion! For a time Hotspur and Douglas carried all before them, dispersing the King's knights and beating his standard to the ground. Then Hotspur was killed by a random arrow that pierced his brain; his death had a disastrous effect upon the morale of his men, who began to disperse in all directions. King Henry advanced all along the line to gain a total victory, as his cavalry pursued Hotspur's disordered followers for some 3 miles. Douglas and numerous other rebel leaders were captured, and the northern army lost 200 knights and esquires, and some 4,000 yeomen and archers. The Royal army lost 10 knights, many esquires and about 2,000 private soldiers. The wounded of both sides amounted to more than 3,000.

Reconstructing the Battle

The heavy casualties resulting from the large numbers of archers might lower morale to such an extent that the remainder of the forces are prevented from ever engaging each other. However, as both sides are employing approximately equal numbers of archers, a satisfactory balance could prevail.

An interesting aspect of this battle is the manner in which Hotspur, Douglas and their knights endeavoured to kill the King, though frustrated by the number of knights wearing the Royal colours. This 'struggle within a struggle', which could form an essential part of the battle, should ideally be fought under 'Individual Skirmish' rules.

Commanders' Classification

King Henry and Prince Hal did well to rally their men after initial reverses and to concert their attacks but, as their force is numerically stronger, classifying either of them as Above Average would weigh the odds too much in their favour. The leaders on both sides, therefore, are all considered to be Average.

Number and Quality of Men

King Henry was marching north with an army to invade Scotland, so that the nucleus of his force at Shrewsbury must have been regular troops. It can be assumed that Henry's force consisted of about 1,500 mounted knights and men-at-arms, 5,000

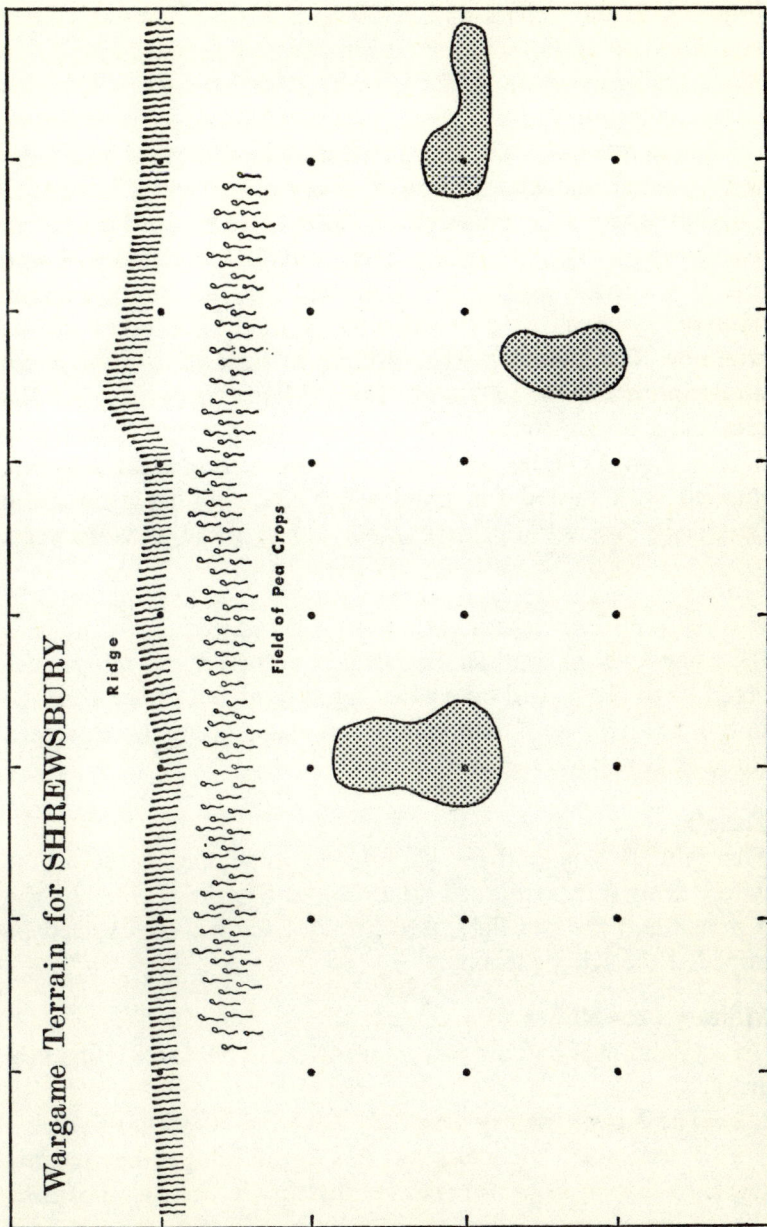

Wargame Terrain for SHREWSBURY

Ridge

Field of Pea Crops

H

infantry, 2,000 archers and 3,500 local levies. Hotspur's force
could consist of 1,000 mounted knights and men-at-arms, 1,000
Scots (250 mounted), 4,000 infantry, 2,000 archers and 2,000
northern levies.

The mounted knights, men-at-arms and infantry on both sides
were trained and experienced professional soldiers. Hotspur's
Cheshire archers should be classed as elite troops, and the Royal
archers as average. Hotspur's levies probably marched with him
out of a certain loyalty to his cause, whereas Henry's levies,
gathered as he marched to Shrewsbury, might not have come so
willingly. The levies probably fought in their usual style as an
undisciplined mob, ill armed and perhaps timorous, but the
remainder of the troops on both sides fought well in the manner
of the day. Douglas and his Scots were first-class soldiers,
imbued with hatred for the English and fighting even more
fanatically by reason of their capture during the previous year.

Morale
It is reasonable to assume that both sides started the battle with
the same level of morale. The usual fluctuations in morale decreed by the rules will take effect upon the King's men after the
initial fire-fight and then, more seriously, upon the northern
force at Hotspur's death.

Terrain
The ground was perhaps slightly advantageous to Hotspur,
whose army was positioned on a ridge and perhaps aided by the
3 ponds and the pea field, which may have inconvenienced or
impeded the King's troops.

Military Possibilities
1   Glendower and his Welshmen arrive to reinforce Hotspur's
army.
2   The King's men do not rally after the initial fire-fight.
3   Prince Hal's tactical sense has not developed beyond his
years, so that he does not lead his division in their vital attack.
4   King Henry is killed.
5   Hotspur is not killed.

14
The Battle of Verneuil
1424

A RELATIVELY rare occurrence for the time, this was no chance
clash but an ordered and premeditated battle, the first for nearly
9 years. The later stages of the Hundred Years War had been
reached. Henry V had died in 1422 and the English were now
commanded by the Regent, John, Duke of Bedford, supported
by such legendary leaders as the Earls of Salisbury and Suffolk,
Lord Scales, Sir John Fastolf and John, Lord Talbot. The
French had been strengthened by a complete Scots army of 6,500
men, besides mercenaries raised in Lombardy and elsewhere.

After the French had tricked the defenders of Verneuil into
surrendering in August 1424, Bedford mustered an army of
8,000 to 9,000 men and at midday on 17 August arrived on the
plain before the town, where the French were drawn up on a
1,000yd front along the crest of a slight ridge that sloped very
gently downwards for 600yd to the forest behind the English
position. Commanded by Count d'Aumale, the force consisted
of 16,000 to 18,000 men, with the Scots division on the right of
the French. Each division formed up in 3 lines, but these merged
into one when the battle began. They had not formed up without
some confusion, as their contingents, drawn from a wide area,
spoke French, English, Italian and a Scots version of English.
The Scottish leaders quarrelled over precedence, eventually
being marched to the forming-up place by the Earl of Buchan,
Constable of France, who then resigned the command to his
father-in-law, the Earl of Douglas.

The allied army was dismounted except for a body of about

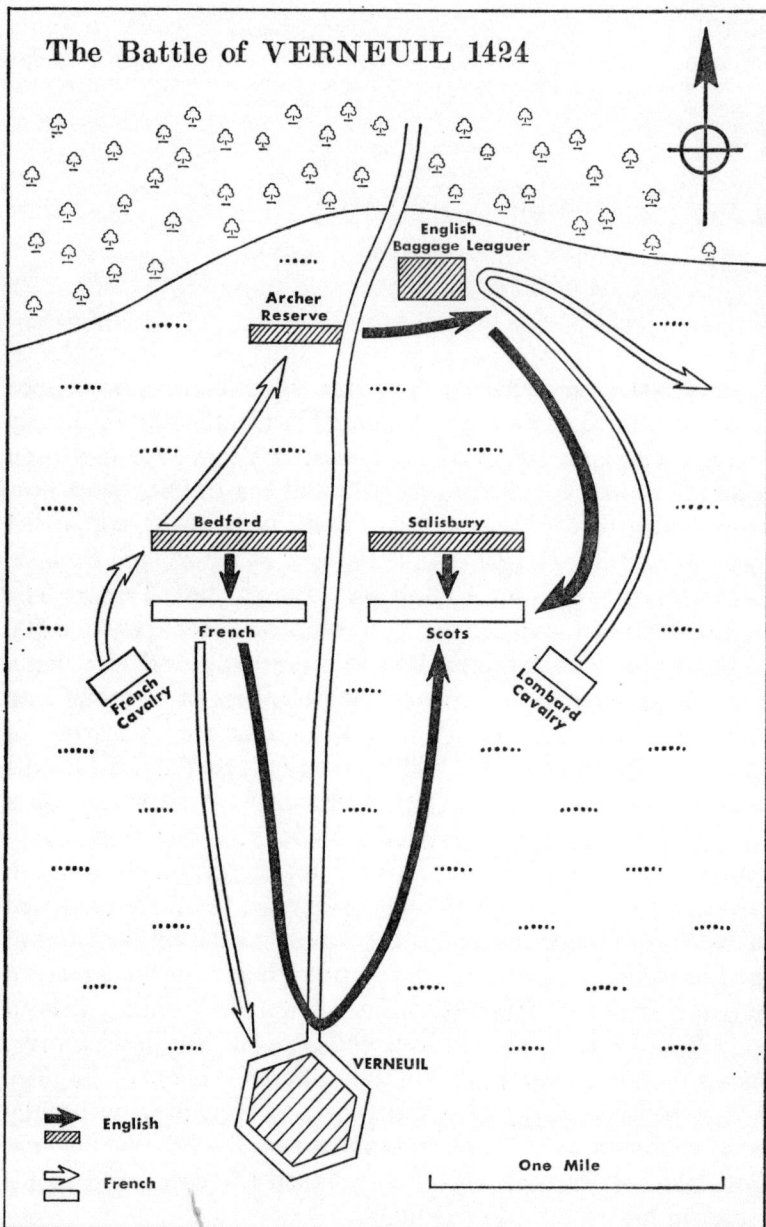

The Battle of VERNEUIL 1424

English Baggage Leaguer

Archer Reserve

Bedford   Salisbury

French   Scots

French Cavalry   Lombard Cavalry

VERNEUIL

English

French

One Mile

900 mounted and armoured Lombard crossbowmen on the right wing, and on the left 1,000 men-at-arms completely mailed, armed with lance and battleaxe and riding armoured horses. The formations included some militia and peasant levies – untrained and ill-armed troops – and some foot crossbowmen were interspersed with the men-at-arms, though little is heard of them during the battle.

After Bedford had marched his army on to the plain, he halted outside missile range, and deployed his men parallel to the enemy and on the same frontage. A conventional soldier, Bedford conformed to the well tried and successful methods of his great-grandfather, Edward III, besides closely following his brother's dispositions at Agincourt. His men formed up in 2 divisions, with dismounted men-at-arms in the centre and archers on the flanks of each division. Salisbury commanded the left and Bedford the right. As a mobile reserve, Bedford stationed 2,000 archers some ¾ mile to the rear, on the right of the road running from Verneuil, and, remembering what had happened to the baggage train at Agincourt, drew his wagons up in a leaguer, with horses tethered head-to-tail in pairs, so that the pages and varlets, having no horses to hold, could use their weapons in defence of the baggage.

Before the battle started, the Scots' leader, Douglas, announced that his men would not give quarter, nor did they expect to receive it.

About 4 o'clock Bedford gave the traditional order 'Avaunt banners!' After kneeling down and reverently kissing the ground, his men responded with the traditional and alarming shout 'St George! Bedford!' before advancing slowly and deliberately. In contrast the French and Scots advanced impetuously and raggedly, the latter probably because of their youth and inexperience. When the English archers were within 250yd of the enemy, they attempted to pass their pointed stakes forward and plant them in the ground, but it was so hard under the August sun that time was wasted trying to hammer them in. Suddenly the body of mailed horse on the French left flank charged the archers and, forcing a passage through the half-erected stakes, rode them down. The survivors gathered together in close formation for mutual protection and, after surging around them, the French horsemen passed on, clattering towards the reserve. This successful attack exposed the right flank of Bedford's division, but it continued to move steadily forward to clash with the French men-at-arms of d'Aumale's division in a hand-to-hand combat that lasted about an hour and was considered, by men who had been at both battles, to be fiercer than at Agincourt. Although outnumbered 2 to 1, the English fought well and, with Bedford prominently wielding his two-handed axe, gradually forced their opponents back, until they turned and fled for the sheltering walls of Verneuil. The triumphant English pursued them as far as the town ditch, where many of them were drowned, and d'Aumale himself was killed before Bedford had collected his scattered troops and begun leading them back to the battlefield.

Salisbury's division encountered much stiffer resistance from the Scots, who resolutely battled with sword, mace and battleaxe, refusing to be dismayed when their French allies broke and fled the field. Although young and untried, the Scots were picked volunteers and excellent fighters.

The Lombard mercenaries on the Scots' right were prevented from helping them because the fire of Salisbury's flank archers drove them wide, where, to their immediate front, they could see the baggage leaguer, unguarded by the reserve archers, who were fiercely engaged with the French cavalry from the other wing. The mounted Lombards therefore swung round the English left and charged into the wagon leaguer, cutting down almost all the poorly armed pages and varlets who tried to defend themselves. Then they pillaged the wagons and collected some of the tethered horses.

The English archers, who had now routed the French mailed cavalry, rushed across to fight off the Lombards, driving them helter-skelter from the field. Having disposed of 2 bodies of cavalry, the exhilarated English archers sought fresh fields to conquer, and, forming up to fearsome shouts, advanced round Salisbury's left and wheeled into the exposed right flank of the sorely tried Scottish division. Meanwhile Bedford's men-at-arms had re-formed and trudged back into the fight – no mean feat for heavily armoured men in the heat of a summer's day. They plunged into the rear of the Scots who, having forsaken quarter, fought to the last man. The enemy lost at least 5,000 men, mainly Scots, and although the majority of the French rank-and-file escaped, most of their leaders were either killed or captured. The English losses were heavy, amounting to about 1,000, but this 'second Agincourt' left the French disheartened and leaderless.

Reconstructing the Battle
Wargames rules are formulated to cater for the norm, but military history is dotted with extraordinary feats of arms, glorious victories against seemingly overwhelming odds, which, far from being regretfully pushed aside as unworkable, have more than earned the right to be reproduced on the wargames table. Victories against odds do not occur by accident. They can be understood by careful analysis of battles and the behaviour of participants. Then, when the reasons for such victories have been discovered, rules can be amended and adapted so as to allow the wargames table to provide a tangible memorial to brave men who have made their mark in the pages of military history.

The English army won the Battle of Verneuil in the face of great numerical odds because its soldiers excelled themselves, lightly armed common soldiers rising to the occasion to defeat more heavily armed men. This can be reproduced on the wargames table without bending the rules in England's favour. Undoubtedly the English army, riding on the crest of the wave of Agincourt, was an exceptional combination of arrogance and good leadership, with a morale well above average. Conversely, the French force suffered from its composition – 2 distinct and ill-coordinated allied forces with a plethora of foreign mercenaries. The whole formed a badly led army of entities, each man aware that the French had not defeated the English in a pitched battle in nearly 100 years of warfare.

In an accurate reconstruction of Verneuil Bedford's division has to defeat twice its own number of men, and the English archers drive off both the French mailed cavalry and the Lombard mounted mercenaries. This can only be done by first-class troops with above average morale opposing second-class troops with below average morale. The Scots division of 6,500 men put up a good fight against Salisbury's division of about 2,000 men, probably aided by the 500 archers posted on his flank, but the Scots, tough and courageous as they were, must suffer from inconclusive leadership and lower fighting qualities, because of their youth and inexperience, than the English, or they would have made their numbers pay.

Commanders' Classification

Bedford and Salisbury, without a doubt, are Above Average commanders. Similarly, there is little argument against their opposite numbers, d'Aumale and Douglas, being Below Average leaders on the day, as neither exercised any tactical control or influence on the battle.

Number and Quality of Men

The French force probably comprised 1,000 French mounted men-at-arms, 900 mounted Lombard crossbowmen, 6,000 French infantry, 500 Genoese crossbowmen, 2,000 French militia and 6,500 Scots.

Bedford's force probably consisted of his own division of 3,000 men, Salisbury's division of 2,000, plus 2,000 archers and an archer reserve of 2,000. He also had an unknown number of poorly armed pages and varlets with the baggage train.

The infantry on both sides, and the French mounted men-at-arms, were trained soldiers, well armed and equipped. The Lombards were professional soldiers, apparently sufficiently skilled in their use of the crossbow to be worthy of their hire, although there is little record of their ever using that weapon at Verneuil. Similarly the foot crossbowmen, probably Genoese, were as ineffective here as at Crécy and other battles of the Hundred Years War. The Scots, probably well-armed pikemen,

Wargame Terrain for VERNEUIL

Slightly Undulating Ground

Baggage Leaguer

were young soldiers. The militia on the French side were typical of the local levies of the medieval period – ill trained, undisciplined, terror-stricken peasants, unwillingly present at a battle from which they took the first opportunity to desert. The aggressive spirit and professional competence of the English archer and man-at-arms are too well known for elaboration.

Morale
As we have said, the English morale was first class. The morale of the French, Scots and mercenaries, affected by lack of common nationality and a background of defeats, was of a lower standard.

Terrain
Fought on a perfectly flat plain, the terrain had no effect whatsoever upon the course of the battle or its outcome.

Military Possibilities
1   The allied French and Scots move sufficiently quickly to catch the English deploying on the plain.
2   The Allies cooperate better and support each other.
3   The French cavalry charges Bedford's exposed right flank instead of attacking the reserve.
4   The Lombards ignore the wagons and attack at a point where they would have been more effective. It is possible, however, that they were 'carried' towards the baggage leaguer by horses excited by the noise of battle.
5   The French cavalry defeats the archer reserve, as they should have done, and then sweeps round on to the rear of Salisbury's division while it is occupied frontally by the Scots.

15

The First Battle of St Albans
1455

IN MAY 1455 Richard, Duke of York, marched south with a
force of some 3,000 men. His army was principally composed of
his personal retainers but later it was augmented by followers of
the Earls of Salisbury and Warwick and Lord Cobham. To
oppose York, King Henry VI hastily collected an army of about
2,000 men, including a quarter or so of the nobility of the land,
among them the Duke of Somerset, York's bitterest enemy. The
2 forces came upon each other near St Albans, which the King's
force reached first, so that York's had to halt just outside the
town.

St Albans was encircled by the remains of a defensive ditch,
and the King's men lined its eastern face where it was crossed by
3 roads – Cock Lane, Shropshire Lane and Sopwell Lane. Each
of them had wooden barriers that could be swung across to
form an obstacle to horsemen or vehicles. The King's soldiers
strengthened these barriers with carts, barrels and similar
obstacles, but left their positions during a fruitless pre-battle
parley, so that they had to reoccupy them hastily when the
Duke of York's army suddenly advanced. York attacked along
the 2 southern roads, coming to an abrupt halt at the barricades,
where, in the subsequent hand-to-hand fighting, the King's men
repulsed all his efforts to break through.

At this point Warwick realised that the King's forces were
divided into 2 by the houses and gardens known as the Town
Backsides, lining Holywell Street. He led his men forward,
therefore, through the gardens and into the houses in this area,

123

The Battle of ST ALBANS 1455

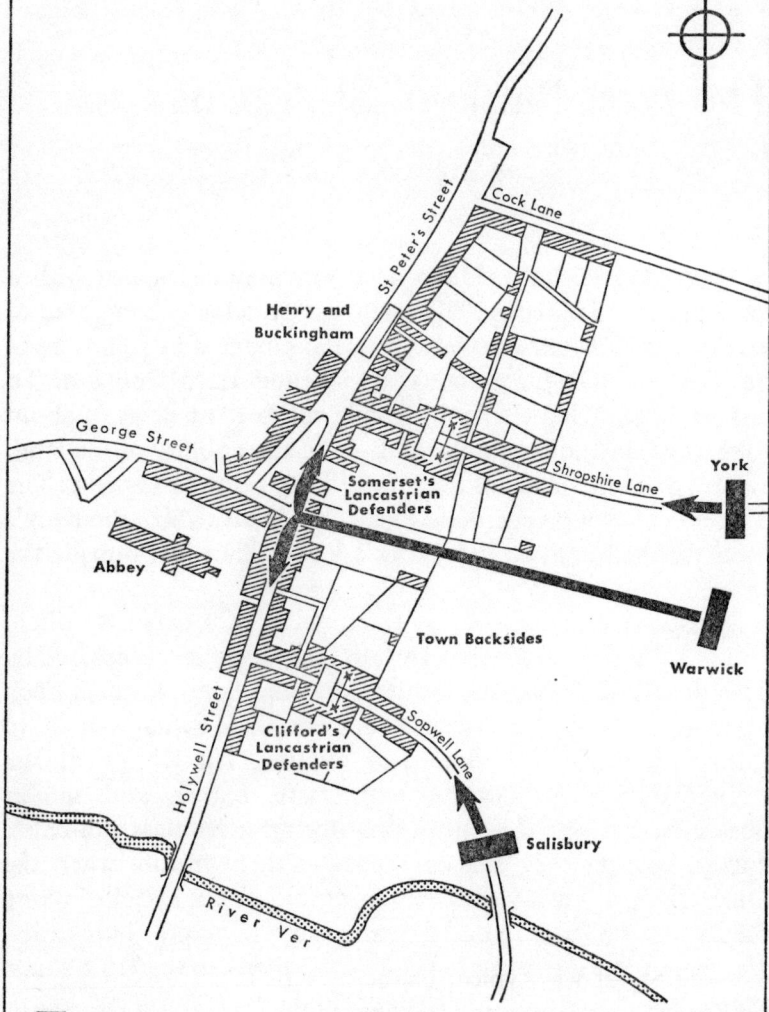

St Peter's Street

Cock Lane

Henry and
Buckingham

George Street

Shropshire Lane

York

Somerset's
Lancastrian
Defenders

Abbey

Town Backsides

Warwick

Holywell Street

Sopwell Lane

Clifford's
Lancastrian
Defenders

Salisbury

River Ver

■ Yorkists
□ Lancastrians
x——x Barricaded Bars

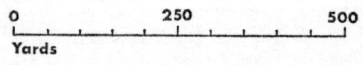

0   250   500
Yards

and, by making holes in the house walls, broke into Holywell Street. There were no Royal defenders here, and Warwick was able, by wheeling right and left, to hit the flanks of the King's forces defending the barricades. The unexpected onslaught caused them to fall back, allowing the whole Yorkist army to burst into the town, and a vast mêlée ensued around the Royal Standard, set up in St Peter's Street. The King's outnumbered troops were dismayed at being forced to abandon defences they seemed capable of holding in the face of all enemy attacks.

The sharp struggle went on for about an hour until the King was wounded by an arrow at close range, causing the Royal Standard bearer to throw down the Standard and flee. At this, panic set in and the remaining Royalist soldiers turned and ran. Largely because Warwick had given orders to 'spare the rank and file and smite only the leaders', not more than 100 men were killed, but most of them were of high rank, including Somerset himself. It is possible that the knights suffered heavy losses because they fought on foot in armour and could not run away when the day went against them. The unimpeded bowmen and billmen, on the other hand, could easily take to their heels.

Reconstructing the Battle
Divide the King's force into 2 bodies at the barricades in Shropshire Lane and Sopwell Lane, and allow the battle to proceed as it did in reality. It will be interesting if the wargamer commanding the King's force is unaware of the true course of events at St Albans in 1455.

Commanders' Classification
Warwick was Above Average, and the rest of the leaders on both sides Average. The failure to garrison the houses in Holywell Street was typical of the lack of control exercised in that day and age, although this was the first battle of the Wars of the Roses. Also, with a force only 2,000 strong and 2 barricades to defend, there could have been few troops to spare, and the manner in which the King's men, having left their positions, had to scramble back when the battle started was perhaps the

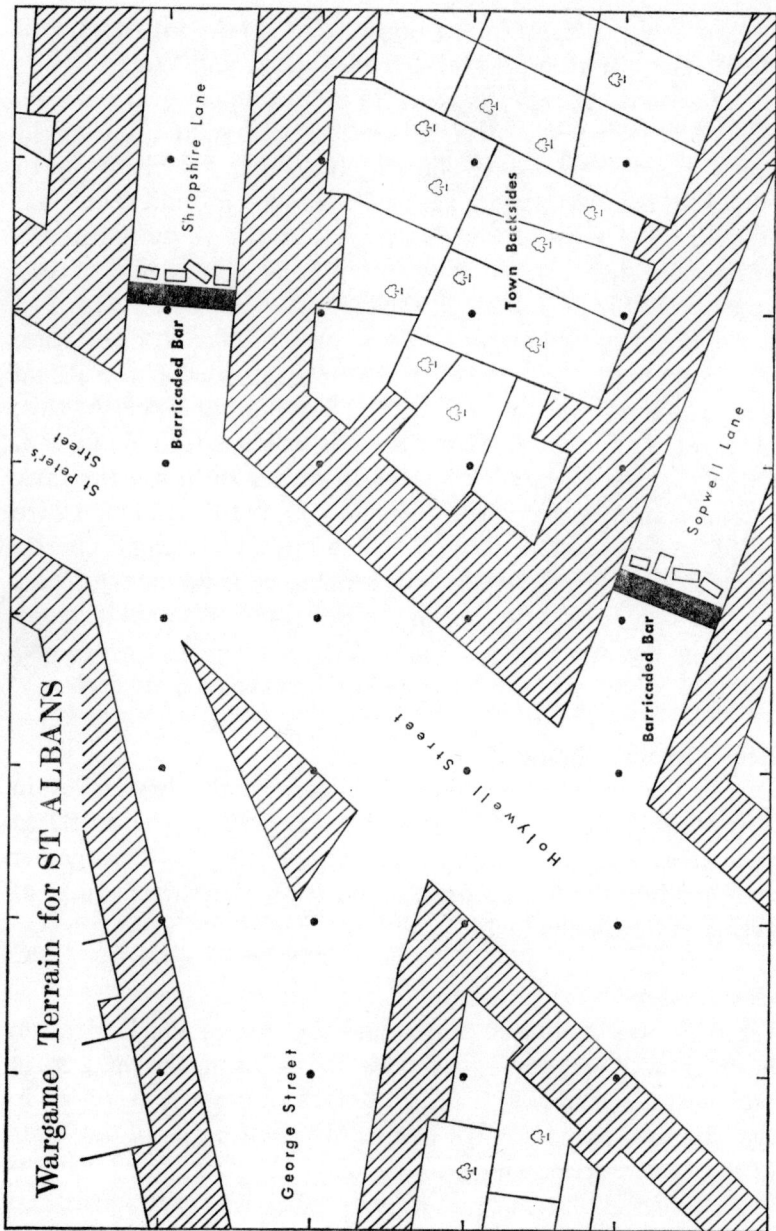

Wargame Terrain for ST ALBANS

St Peter's Street

Shropshire Lane

Barricaded Bar

Town Backsides

Sopwell Lane

Barricaded Bar

Holywell Street

George Street

reason for no one assuming responsibility for the apparently solid barrier of houses stretching between the 2 barricades.

Number and Quality of Men
Both forces consisted entirely of 'gentlemen', their retainers and their small private armies of archers and billmen. The King's army of about 2,000 probably included 350–500 archers, while the Duke of York's force possibly had 500–750 of them.

It is recorded that a large proportion of the combatants on both sides had fought in France and were fairly well trained and experienced – certainly as good as could be expected at the start of a civil war. The foot soldiers were armed with pikes, bills and axes, and the gentry wore armour but fought on foot. Neither side used field artillery or handguns at this battle.

Morale
Both sides were probably of average morale until Warwick attacked the Royalist barricades in flank, when the defenders' morale must have fallen.

Terrain
The terrain in this battle was all important, Warwick using it to make his way unseen into the enemy line. The barricades formed focal points for the fighting.

Military Possibilities
1 The Lancastrians man their defences more thoroughly, although 2,000 men is far too few to maintain defensive positions on each flank and also garrison a number of houses.
2 Warwick lacks the perception to break through the houses.

Appendix 1
Rules

THE RULES that control a wargame usually reflect the character and temperament of their devisor: a dashing wargamer will allow full opportunities for panache in glorious charges and heroic defence, and a steady player scope for coolness in a commander and intricate manoeuvre. This individuality among wargamers leads them to adapt sets of rules not of their own devising to their own ideas and military concepts. When doing this, certain facts should be borne in mind if the rules are to provide a brisk and realistic historical reconstruction. Remember first that the little metal and plastic soldiers used in wargaming have only the fighting ability or morale that you bestow upon them. Where history decrees that particular soldiers are known to be of inferior morale and fighting ability, penalising rules must be devised to govern their performance. The battles in this book were fought largely between sturdy professional soldiers of roughly equal ability, although they were not equally well led. Where levies and other inferior troops are involved, however, the rules must reflect their poorer morale and lesser fighting ability.

The battles herein embody such factors as numerical disparity of forces, surprise, weapons of varying power, and differing states of morale. All are difficult to simulate by means of normal wargames rules, so that variants have to be devised. Then there are those almost legendary occasions when a commander takes a deliberately calculated risk that comes off. Such departures from the norm are not easy to re-enact, and the rules have to be slanted to accommodate them.

Commercially produced sets of rules, in ever-increasing

numbers, are regularly advertised in wargamers' and modellers' magazines. One of the best sets for the periods covered in this book are those of the Wargames Research Group (Bob O'Brien, 75 Ardingly Drive, Goring-by-Sea, Sussex), whose compilers claim that 'great weight has been given to the characteristics of the different people that form the warrior types of Ancient armies so as to bring out their varying fighting qualities and reactions to stress'. Other relevant commercial rules are those for the Wars of the Roses, produced in Great Britain by Decalset; and Chainmail, a set of rules for medieval wargaming devised by the Americans Gary Gygax and Jeff Perren (Guidon Games, PO Box 1123, Evansville, Ind 47713, USA). Beginners' rules for both ancient and medieval periods that control a battle in not too complex a fashion have been prepared by Tony Bath, and can be obtained from D. F. Featherstone, 69 Hill Lane, Southampton SO1 5AD, Hants.

Earlier works by D. F. Featherstone (*Wargames*; *Advanced Wargames*; *Wargames Campaigns*; *Wargames through the Ages*, Vol I; *Battles with Model Soldiers*; and *Solo Wargaming*) contain rules, together with advice and suggestions for compiling rules to fit varying circumstances.

The well-known international wargames magazine *Wargamer's Newsletter* (Belmont-Maitland Publishers, Tradition, 188 Piccadilly, London W1) regularly publishes articles on rules, together with features and wargames reports on battles in this period. Another source is the Society of Ancients (John Norris, Hillside, Jackson's Lane, Highgate Village, London N6 5SR), which caters for the wargamer in periods ranging from Ancient Egypt to the Middle Ages. The Society's magazine, *Slingshot*, is a mine of information.

I

Appendix 2
Terrain

IT IS most important in the reconstruction of a historical battle that the terrain should closely resemble the actual battlefield, though only the areas fought over need be reproduced. Marches bringing the armies into contact may be conducted on the 'surrounding area' map (see p 18). Every battle in this book may be fought over a terrain set out on a playing surface of 8ft × 5ft, and the wargames terrain maps are scaled down to fit this size.

The easiest manner of constructing the undulating surface of a wargames terrain is to stretch a green cloth or suitably coloured plastic sheeting over mounds of books, slabs of polystyrene or pieces of wood placed in position on the table top. Rivers and roads can be represented by strips of Fablon stuck on the cloth or painted on the plastic with poster paints. Trees and hedges may be made from moss or bought ready made from model railway and hobby shops. Stone walls, bridges and rail fencing can be made from balsa wood, and crags and rocky outcrops from broken pieces of polystyrene packing.

Although vegetation might not be shown on a terrain map, few stretches of land are completely bare, and the appearance of any battlefield is greatly improved by scattered clumps of trees, bushes and scrub. Woods must be so constructed that troops can manoeuvre within them, and they are best represented by a few trees fixed to an irregularly shaped piece of hardboard or cardboard. The houses that form the greatest part of the terrain for St Albans can either be made from small blocks of wood, as single dwellings or as a row (or terrace). There are suitable

130

plastic and cardboard cut-out kits to make houses, obtainable from model-railway shops.

Notes on wargames battlefield construction are contained in the booklet *Wargames Terrain* by D. F. Featherstone.

Appendix 3
Availability of Wargames Figures

THE NAMES and addresses listed below are those of the principal makers of wargames figures at the time of writing. The best known of them, such as Miniature Figurines, Hinchliffe, Jack Scruby etc, produce figures for most of the battles covered by this book. All their products are good, and, in time, the wargamer will decide for himself which he prefers.

The wargamer has a considerable choice of manufacturers who make model soldiers for wargames in the following scales: 5mm (regimental blocks), 9mm (N gauge), 15mm, 20mm, 25mm, 30mm (rounds and flats), 40mm, and 54mm. Scales do not necessarily coincide, and the so-called 20mm figure of one manufacturer may appear minute, or even gigantic, compared with a 20mm figure from another. It is essential, therefore, to check figure sizes before buying from different manufacturers, if one's armies are not to be represented by a mixture of giants and pygmies.

Apart from the 5mm Mini-Minifigs regimental blocks, the cheapest possible way of building up wargames armies is by using the inexpensive 20mm plastic figures produced by Airfix Products Ltd, which at present comprise more than 40 different types and periods. It is also quite easy to convert these plastic figures into soldiers of other periods – in fact no type of soldier cannot be so converted from one or other of the existing Airfix production figures. Instructions for converting plastic figures are given in *Military Modelling* by D. F. Featherstone and *How to Go Advanced Plastic Modelling* by Chris Ellis, and illustrated articles on converting appear regularly in *Airfix Magazine, Military Modelling, Slingshot, Wargamer's Newsletter* and other journals.

Of particular interest and value to wargamers specialising in the ancient period is a series of articles, 'Roman Friends and Foes', by Bob O'Brien, published by *Airfix Magazine* (February and November/December 1968; January, February, March, April, May and June 1969). This series not only details the characteristics and tactics of the Romans and their numerous enemies, but also gives explicit instructions on converting standard Airfix figures into numerous other types of warrior, such as Goths, Picts, Ancient Britons, Numidians, Gauls, Germans and Arabs. It also deals with specific types, such as camel archers, armoured-horse archers, slingers, Roman artillery and siege weapons, war elephants, chariots and boats, and is well illustrated with photographs and drawings.

MANUFACTURERS IN GREAT BRITAIN

Airfix Hobby & Toy Sales Ltd (plastic 20 and 54mm)

Britains Ltd (plastic and metal 54mm)

Greenwood & Ball Ltd, 61 Westbury Street, Thornaby-on-Tees, Teesside (metal 25 and 30mm)

Hinchliffe Models Ltd, Meltham, Yorks (metal 25mm)

Hinton Hunt, Rowsley, River Road, Taplow, Bucks (metal 25mm)

Peter Kemplay, Framlingham, Woodbridge, Suffolk (Tradition and Lamming's metal 25mm)

Peter Kirk, 3 Wynfield Road, Western Park, Leicester (25mm metal)

Peter Laing, 11 Bounds Oak Way, Southborough, Tunbridge Wells, Kent TN4 OUB (metal 15mm)

W. H. Lamming, 130 Wexford Avenue, Greatfield, Kingston-upon-Hull, Yorks (metal 25mm)

Leicester Micro-Models, 5 Goss Barton, Nailsea, Bristol BS19 2XD (medieval 1:300 scale)

Miniature Figurines, 28–32 Northam Road, Southampton, Hants (metal 5, 16, 25 and 30mm)

Model Figures & Hobbies, 8 College Square North, Belfast BT1 7HS (Segom 25mm plastic)

The Northern Garrison, Castlegate, Knaresborough, Yorks (metal 20, 25 and 30mm)

Phoenix Model Developments Ltd, The Square, Earls Barton, Northampton (metal 20, 25 and 30mm)

Rose Miniatures, 15 Llanover Road, Plumstead, London SE18 3ST (metal 20 and 25mm)

Spencer-Smith Miniatures, 66 Longmeadow, Frimley, Camberley, Surrey (30mm plastic)

Tradition, 188 Piccadilly, London W1 (metal 25 and 30mm)

Warrior, 23 Grove Road, Leighton Buzzard, Beds LU7 8SF (metal 20 and 33mm)

Willie Figures, 60 Lower Sloane Street, London SW3 (metal 30mm)

USA

Bugle & Guidon, PO Box 662, Jackson, Michigan 49204 (metal 30mm)

Command Post, 760 West Emerson, Upland, California 91786 (metal 30mm)

Hobby House, 9622 Ft Meade Road, Laurel, Maryland 20810 (metal 20mm)

C. H. Johnson, PO Box 281, Asbury Park, New Jersey 07713 (metal 20, 25 and 30mm)

Jack Scruby's Military Miniatures, PO Box 1658, Cambria, California 93428 (metal 9, 20, 25, 30 and 40mm)

K. & L. Company, 1929 North Beard, Shawnee, Oklahoma 74801 (metal 20mm)

Der Kriegspieler, PO Box 419, Bedford, Mass (metal 20mm)

SPAIN

Alymers Miniploms, Maestro Lope 7, Burjasot (metal 20mm)

SWEDEN

Holgar Eriksson, Sommarrovagen 8, Karlstad (metal 30mm)

FRANCE

Segom, 50 Boulevard Malesherbes, Paris 8. UK Distributors – Model Figures & Hobbies, Belfast (plastic 30mm)

Starlux. UK Distributors – Beatties Ltd, 112 High Holborn, London WC1 (plastic 30mm)

WEST GERMANY

Elastolin, Neustadt/Coburg, Jahreswende (plastic 40mm and 60mm)

Walter Merten, 1 Berlin 42 Tempelhof, Industriestrasse 25 (plastic 30 and 45mm)

'Flats' are two-dimensional 30mm figures stamped out of thin metal, and it is possible to obtain any type of soldier, of any period of history and in any position from one or other of these suppliers:

Alloys Ochel, Feldstrasse 24b, Kiel, West Germany

Rudolph Donath, Schliessfath 18, Simbach/Inn OBB, West Germany

F. Nichel, Goethestrasse 16, Wendlingen am Neckar, West Germany

Bibliography

THE FOLLOWING books were used when researching the battles, and many of them cover more than one:

Barrett, C. R. B. *Battles and Battlefields in England* (1896): Stamford Bridge, Lewes, Neville's Cross, Shrewsbury and St Albans

Burne, Lt-Colonel A. H. *The Crécy War* (1955): Bannockburn, Morlaix and Auberoche

Cambridge Ancient History, Vol III: Kadesh, Ticinus River and Cynoscephalae

Dupuy, R. E. and T. N. *The Encyclopaedia of Military History from 3500 BC to the Present* (1970): Leuctra, River Jaxartes, Cynoscephalae and Taginae

Featherstone, D. F. *The Bowmen of England* (1967): Bannockburn, Morlaix, Neville's Cross and Verneuil

——. *Wargames through the Ages*, Vol I (1972): Kadesh, Leuctra, Taginae and Stamford Bridge

Grant, James. *British Battles* (nd): Stamford Bridge, Bannockburn, Shrewsbury and St Albans

Green, Lt-Colonel Howard. *Guide to the Battlefields of Britain and Ireland* (1973): Stamford Bridge, Lewes, Neville's Cross and Shrewsbury

Kinross, John. *Discovering Battlefields in Northern England and Scotland* (1968): Stamford Bridge, Lewes, Neville's Cross, Shrewsbury and St Albans

Montross, Lynn. *War through the Ages* (1944): Leuctra, River Jaxartes, Cynoscephalae and Taginae

Treece, Henry and Oakeshott, Ewart. *Fighting Men* (1963): Taginae and Bannockburn

Wood, Sir Evelyn. *British Battles on Land and Sea* (1915):

Stamford Bridge, Bannockburn, Shrewsbury and St Albans

KADESH

Burne, Lt-Colonel A. H. *The Art of War on Land* (1944)
Montgomery, Field-Marshal Viscount. *A History of Warfare* (1968)

LEUCTRA

Eggenberger, David. *A Dictionary of Battles from 1479 BC to the Present* (1967)

RIVER JAXARTES

Dupuy, T. N. *The Military Life of Alexander the Great of Macedon* (1969)
Lamb, Harold. *Alexander of Macedon* (1946)

TICINUS RIVER

Cottrell, L. *An Enemy of Rome* (1962)
Dupuy, T. N. *The Military Life of Hannibal* (1969)

CYNOSCEPHALAE

Birnie, Arthur. *The Art of War* (1942)
Pratt, Fletcher. *The Battles That Changed History* (1956)

TAGINAE

Barker, Phil. *The Armies and Enemies of Imperial Rome* (1972)
Coggins, Jack. *The Fighting Man* (1966)
Graves, Robert. *Count Belisarius* (1938)

STAMFORD BRIDGE

Butler, Denis. *1066 – The Story of a Year* (1966)
Furneaux, Rupert. *Conquest 1066* (1966)

VERNEUIL

Burne, Lt-Colonel A. H. *The Agincourt War* (1956)

The magazines *Wargamer's Newsletter*, *Slingshot* and *Tradition* also contributed much material of value.

These books are also recommended:

Boudet, J. *The Ancient Art of Warfare* (1969)
Chandler, D. *The Art of Warfare on Land* (1974)
Fuller, J. F. C. *The Generalship of Alexander the Great* (1958)
Halevy, D. *Armies and Their Arms* (1962)
Norman, A. V. B. and Pottinger, D. *Warrior to Soldier 449–1660* (1966)
Oman, C. W. C. *The Art of War in the Middle Ages, AD 378–1515* (1953)
Perroy, E. *The Hundred Years War* (1951)
Turney-High, H. H. *Primitive War* (1949)
Weller, J. *Weapons and Tactics* (1966)
Yadin, Y. *The Art of Warfare in Biblical Lands* (1960)

Here are the best known books in the fast growing list of wargames literature:

Featherstone, D. F. *Wargames* (1962)
——. *Naval Wargames* (1966)
——. *Air Wargames* (1967)
——. *Advanced Wargames* (1969)
——. *Battles with Model Soldiers* (1970)
——. *Wargames Campaigns* (1970)
——. *Solo Wargames* (1972)
——. *Battle Notes for Wargamers* (1973)
——. *Wargames through the Ages*, Vol I, *Ancient and Mediaeval Periods* (1973); Vol II, *1420–1783* (1974); Vol III, *1792-1859* (1975)
——. *Tank Battles in Miniature – The Western Desert Campaign* (1973)
Grant, Charles. *Battle! Practical Wargaming (World War II)* (1970)
——. *The Wargame* (1971)
——. *Napoleonic Wargaming* (1973)
Morschauser, J. *Wargames in Miniature* (1963)
Tunstill, John. *Discovering Wargaming* (1969)
Wells, H. G. *Little Wars* (1913)

Wise, Terence. *Introduction to Battle Gaming* (1969)
Young, Brigadier P. (ed). *The Wargame* (1972)
Young, Brigadier P. and Lawford, Lt-Colonel J. P. *Charge!*
  (1957)

The following books will help the wargamer in assembling his
miniature armies:

Baldet, M. *Lead Soldiers and Figurines* (1961)
Bard, Bob. *Making and Collecting Military Miniatures* (1959)
Featherstone, D. F. *Tackle Model Soldiers This Way* (1965)
——. *Handbook for Model Soldier Collectors* (1969)
——. *Military Modelling* (1970)
Garratt, John G. *Model Soldiers: A Collector's Guide* (1960)
Harris, H. *Model Soldiers* (1962)
——. *How to Go Model Soldier Collecting* (1969)
Nicollier, Jean. *Collecting Toy Soldiers* (1967)
Risley, C. and Imrie, W. *Model Soldiers' Guide* (1964)

Index

141